Solutions Manual to Accompany
Applied Logistic Regression
Second Edition

Solutions Manual to Accompany
Applied Logistic Regression
Second Edition

DAVID W. HOSMER, Jr.
STANLEY LEMESHOW

Solutions Manual Authored by
ELIZABETH DONOHOE COOK

A Wiley-Interscience Publication
JOHN WILEY & SONS, INC.
New York · Chichester · Weinheim · Brisbane · Singapore · Toronto

This text is printed on acid-free paper. ⊗

Copyright © 2001 by John Wiley & Sons, Inc. All rights reserved.

Published simultaneously in Canada.

No part of this publication may be reproduced, stored in a retrieval system or transmitted in any form or by any means, electronic, mechanical, photocopying, recording, scanning or otherwise, except as permitted under Section 107 or 108 of the 1976 United States Copyright Act, without either the prior written permission of the Publisher, or authorization through payment of the appropriate per-copy fee to the Copyright Clearance Center, 222 Rosewood Drive, Danvers, MA 01923, (978) 750-8400, fax (978) 750-4744. Requests to the Publisher for permission should be addressed to the Permissions Department, John Wiley & Sons, Inc., 605 Third Avenue, New York, NY 10158-0012, (212) 850-6011, fax (212) 850-6008, E-Mail: PERMREQ @ WILEY.COM.

For ordering and customer service, call 1-800-CALL WILEY.

Library of Congress Cataloging in Publication

ISBN 0-471-20826-4 (paper)

Printed in the United States of America

10 9 8 7 6 5 4 3 2 1

CONTENTS

PREFACE

In the 10 years between the first and second editions of *Applied Logistic Regression* many readers and instructors asked about the availability of a solutions manual. We recognized the need but quite simply did not have the time to devote to producing one.

The problems in the book are designed to illustrate the methods, ideas and approaches to analysis described in each chapter. Each chapter has relatively few problems. There are no "math" problems in the sense of having to derive a mathematical solution to a theoretical exercise. As such, the problems do not lend themselves to providing a partial solutions manual for students and a complete solutions manual for instructors. We suggest that instructors supplement the problems in the text with similar questions using data from their own fields and areas of statistical practice. We are always interested in seeing new data that provide good teaching examples and sample data sets for additional exercises and exams. If you have such a data set please contact us and we will discuss including it in the archive of statistical data sets at the University of Massachusetts

(http://www-unix.oit.umass.edu/~statdata).

This solutions manual presents the methods, computer output and discussion that we would use if we had been assigned the problems in the text. In any data analysis exercise one makes choices along the way, for example, which variables to include and how to scale continuous covariates. There is both art and science in a good data analysis and two experienced analysts may arrive at slightly different models, each of which accomplishes the goals of the analysis. Thus in many problems our solution should not be taken as the only possible one. We encourage instructors to consider alternative solutions and models and to discuss their respective strengths and weaknesses with their students.

We performed most of the calculations in this solutions manual using STATA (versions 6.0 and 7.0). In several instances SAS was used. The code presented in the manual is what we used to get the job done and likely does not represent the most elegant or efficient coding. We have made no attempt to use all the "latest" features in the software. In addition, we have no plans to revise the solutions manual to illustrate future software developments. Virtually all the calculations performed in STATA can be performed in other packages.

We have made every attempt to make the solutions as accurate as possible. There is a formidable amount of numerical computation and calculation in the manual and there are likely a few errors we missed. We would appreciate learning of these. We do not, however, have the time to react to and comment on alternative solutions to the problems.

As noted in the Preface of the text all data sets may be found at the Wiley ftp site,

ftp://ftp.wiley.com/public/sci_tech_med/logistic,

or on the data set archive whose URL is shown above.

We very much appreciate the careful reading and suggested changes Gabriel Suciu and Meng Chen made to this solutions manual.

DAVID W. HOSMER, JR.
STANLEY LEMESHOW
ELIZABETH DONOHOE COOK

Chapter One – Solutions

1. *In the ICU data described in Section 1.6.1, the primary outcome variable is vital status at hospital discharge, STA. Clinicians associated with the study felt that a key determinant of survival was the patient's age at admission, AGE.*

   (a) *Write down the equation for the logistic regression model of STA on AGE. Write down the equation for the logit transformation of this logistic regression model. What characteristic of the outcome variable, STA, leads us to consider the logistic regression model as opposed to the usual linear regression model to describe the relationship between STA and AGE?*

   Logistic regression model:

   $$\pi(AGE) = \frac{e^{\beta_0 + \beta_1 *(AGE)}}{1 + e^{\beta_0 + \beta_1 *(AGE)}}$$

   Logit transformation:

   $$g(AGE) = \beta_0 + \beta_1 * (AGE)$$

   We consider the logistic regression model, rather than the usual linear regression model to describe the relationship between STA and AGE because the outcome variable, STA, is dichotomous, taking on the values 0 and 1.

   (b) *Form a scatterplot of STA versus AGE.*

. **graph sta age**

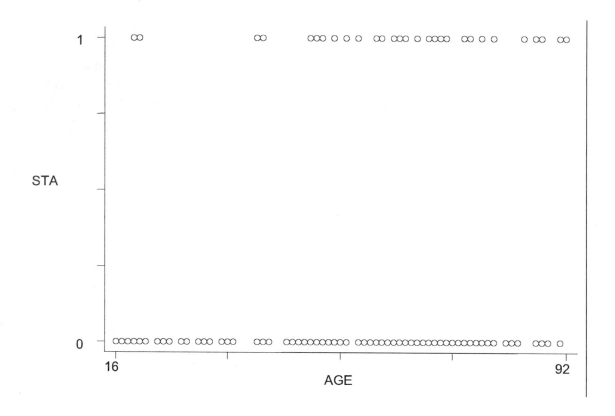

(c)   Using the intervals [15,24], [25,34], [35,44], [45, 54], [55, 64], [65, 74], [75,84], [85, 94] for AGE, compute the STA mean over subjects within each AGE interval. Plot these values of mean STA versus the midpoint of the AGE interval using the same set of axes as was used in Exercise 1(b).

AGEGP	Interval	Midpoint	n	Mean STA
1	15-24	20.0	26	0.077
2	25-34	30.0	8	0
3	35-44	40.0	11	0.182
4	45-54	50.0	25	0.200
5	55-64	60.0	39	0.205
6	65-74	70.0	50	0.180
7	75-84	80.0	30	0.300
8	85-94	90.0	11	0.455

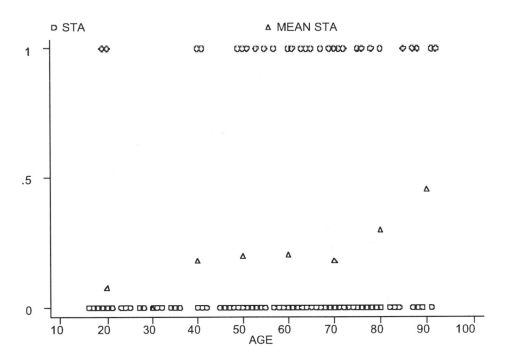

(d)   *Write down an expression for the likelihood and log likelihood for the logistic regression model in Exercise 1(a) using the ungrouped, n=200, data. Obtain expression for the two likelihood equations.*

likelihood function:

$$l(\beta) = \prod_{i=1}^{n} \zeta(x_i) \qquad \text{where, } \zeta(x_i) = \pi(x_i)^{y_i}\left[1 - \pi(x_i)\right]^{1-y_i}$$

$$x = \text{AGE} \qquad \text{and } y_i = \begin{cases} 0 & \text{if the patient lived} \\ 1 & \text{if the patient died} \end{cases}$$

log likelihood function:

$$L(\beta) = \ln\left[l(\beta)\right] = \sum_{i=1}^{n} \left\{ y_i \ln\left[\pi(x_i)\right] + (1 - y_i)\ln\left[1 - \pi(x_i)\right] \right\}$$

likelihood equations:

$$\sum_{i=1}^{n}\left[y_i - \pi(x_i)\right] = 0$$

and

$$\sum_{i=1}^{n} x_i \left[y_i - \pi(x_i) \right] = 0$$

(e)   *Using a logistic regression package of your choice, obtain the maximum likelihood estimates of the parameters of the logistic regression model in Exercise 1(a). These estimates should be based on the ungrouped, n=200, data. Using these estimates, write down the equation for the fitted values, that is, the estimated logistic probabilities. Plot the equation for the fitted values on the axes used in the scatterplots in Exercises 1(b) and 1(c).*

```
.  logit sta age

Iteration 0:   Log Likelihood =-100.08048
Iteration 1:   Log Likelihood =-96.288372
Iteration 2:   Log Likelihood =-96.153701
Iteration 3:   Log Likelihood = -96.15319

Logit Estimates                            Number of obs =     200
                                           chi2(1)       =    7.85
                                           Prob > chi2   = 0.0051
Log Likelihood =  -96.15319                Pseudo R2     = 0.0392

------------------------------------------------------------------------
    sta |    Coef.    Std. Err.       z      P>|z|     [95% Conf. Interval]
--------+---------------------------------------------------------------
    age |  .0275426   .0105645     2.607    0.009     .0068366    .0482487
  _cons | -3.058513   .6961091    -4.394    0.000    -4.422862   -1.694165
------------------------------------------------------------------------
```

logistic regression model (fitted values):

$$\pi(AGE) = \frac{e^{-3.058513 + 0.0275426*(AGE)}}{1 + e^{-3.058513 + 0.0275426*(AGE)}}$$

logit transformation:

$$g(AGE) = -3.058513 + 0.0275426*(AGE)$$

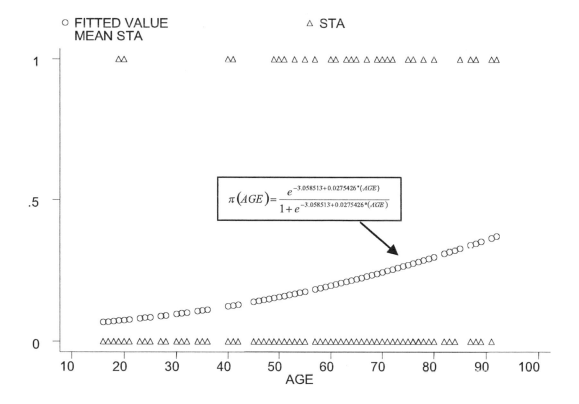

○ FITTED VALUE
MEAN STA

△ STA

(f)   *Summarize (describe in words) the results presented in the plot obtained from Exercises 1(b), 1(c) and 1(e).*

The plot of STA vs. AGE (indicated in the scatterplot in 1(b)) demonstrates the dichotomous nature of the STA variable, which takes on the value zero if a patient is discharged alive or the value one if the patient died prior to discharge. The plot suggests that older people are more likely to die in the ICU, although overall, people are more likely to live than to die.

The plot of mean STA by age group (indicated in the scatterplot in 1(c) by triangles) reinforces this impression. In general, as age increases, the probability of dying in the ICU increases.

The plot of the estimated logistic probabilities vs. AGE (indicated in Exercise 1(e) scatterplot by circles) indicates that the probability of dying does increase with increasing age. The rate of increase in the probabilities seems to increase with increasing age.

(g)   *Using the results of the output from the logistic regression package used for Exercise 1(e), assess the significance of the slope coefficient for AGE using the likelihood ratio test, the Wald test, and, if possible the Score test. What assumptions are needed for the p-values computed for each of these tests to be valid? Are the results of these tests consistent with one another? What is the value of the deviance for the fitted model?*

Deviance:

$$D = -2 \ln \left[\frac{\text{(likelihood of the current model)}}{\text{(likelihood of the saturated model)}} \right]$$

$$D = -2(-96.15319)$$

$$D = 192.30638$$

Likelihood Ratio Test:

$$H_0: \beta_1 = 0$$

$$H_A: \beta_1 \neq 0$$

$$G = D(\text{model without variable}) - D(\text{model with variable})$$

$$G = 200.16 - 192.31$$

$$G = 7.85 \qquad\qquad G \sim \chi^2(1) \qquad\qquad p = 0.0051$$

$\therefore$ reject H_0, it is not consistent with the data that $\beta_1 = 0$; we conclude that AGE is a significant predictor of STA.

Assumption: the statistic G will follow a χ^2 distribution with 1 degree of freedom under the null hypothesis.

Wald Test:

$$H_0: \beta_1 = 0$$

$$H_A: \beta_1 \neq 0$$

$$W = \frac{\hat{\beta}_1}{S\hat{E}(\hat{\beta}_1)} = 2.607 \quad W \sim N(0,1) \qquad\qquad p = 0.009$$

$\therefore$ reject H_0, it is not consistent with the data that $\beta_1 = 0$; we conclude that AGE is a significant predictor of STA.

Assumption: the Wald statistic will follow a normal distribution with mean 0 and variance 1.

Score Test:

$$H_0: \beta_1 = 0$$

$$H_A: \beta_1 \neq 0$$

$$ST = \frac{\sum_{i=1}^{n} x_i (y_i - \bar{y})}{\sqrt{\bar{y}(1 - \bar{y}) \sum_{i=1}^{n} (x_i - \bar{x})^2}}$$

$$ST = \frac{303.2}{\sqrt{0.16(80035.6)}} = \frac{303.2}{113.16} = 2.68 \qquad p = 0.007$$

$$ST \sim N(0,1)$$

$\therefore$ reject H_0, it is not consistent with the data that $\beta_1 = 0$; we conclude that AGE is a significant predictor of STA.

Assumption: The score statistic is normally distributed with mean 0 and variance 1.

The results of the likelihood ratio test, the Wald test and the Score test are consistent. Each test indicates that the model is significant.

The value of the deviance for the fitted model is $D = 192.31$.

(h)   *Using the results from 1(e), compute 95 percent confidence intervals for the slope and constant term. Write a sentence interpreting the confidence interval for the slope.*

```
. logit sta age

Iteration 0:   Log Likelihood =-100.08048
Iteration 1:   Log Likelihood =-96.288372
Iteration 2:   Log Likelihood =-96.153701
Iteration 3:   Log Likelihood = -96.15319

Logit Estimates                                    Number of obs =     200
                                                   chi2(1)       =    7.85
                                                   Prob > chi2   = 0.0051
Log Likelihood =  -96.15319                        Pseudo R2     = 0.0392

-----------------------------------------------------------------------------
    sta |    Coef.    Std. Err.      z      P>|z|    [95% Conf. Interval]
--------+--------------------------------------------------------------------
    age |  .0275426    .0105645    2.607    0.009    .0068366    .0482487
  _cons | -3.058513    .6961091   -4.394    0.000   -4.422862   -1.694165
-----------------------------------------------------------------------------
```

The confidence intervals shown in the STATA output above (6th and 7th columns), can be computed using equations (1.15) and (1.16) from the text.

Endpoints of a 100(1-α)% confidence interval for slope coefficient:

$$\hat{\beta}_1 \pm z_{1-\alpha/2}\hat{SE}\left(\hat{\beta}_1\right)$$

$$0.0275 \pm 1.96(0.0106)$$

$$(0.0068,\ 0.0482)$$

Endpoints of a 100(1-α)% confidence interval for constant:

$$\hat{\beta}_0 \pm z_{1-\alpha/2}\hat{SE}\left(\hat{\beta}_0\right)$$

$$-3.0585 \pm 1.96(0.6961)$$

$$(-4.4229,\ -1.6941)$$

The 95% confidence interval for the slope suggests that the change in the log odds of dying in the ICU (STA=1) per one year increase in AGE is 0.0275 and the change could be as little as 0.0068 or as much as 0.0482 with 95% confidence.

(i)   *Obtain the estimated covariance matrix for the model fit in 1(e). Compute the logit and estimated logistic probability for a 60-year old subject. Compute 95 percent confidence intervals for the logit and estimated logistic probability. Write a sentence or two interpreting the estimated probability and its confidence interval.*

```
. correlate, _coef cov

        |         AGE        _cons
--------+-------------------
    AGE|    .000112
  _cons|   -.007104    .484568
```

$$\hat{g}(60) = -3.0585 + 0.0275(60)$$

$$= -1.4060$$

$$Var\left[\hat{g}(60)\right] = 0.4846 + (60)^2 \times (0.000112) + 2 \times 60 \times (-0.007104)$$

$$= 0.0353$$

$$SE\left[\hat{g}(60)\right] = 0.1879$$

The estimated standard error of the logit for a 60-year old subject, can be used to calculate the endpoints of the confidence interval for the logit for a 60-year old subject:

$$\hat{g}(60) \pm z_{1-\alpha/2} SE\left[\hat{g}(60)\right]$$

$$-1.4060 \pm 1.96 * 0.1879$$

$$(-1.7741, -1.0378)$$

The estimated logit and the endpoints of its confidence interval can be used to obtain the estimated logistic probability and its confidence interval:

$$\hat{\pi}(60) = \frac{e^{\hat{g}(60)}}{1 + e^{\hat{g}(60)}}$$

$$= \frac{e^{-3.059 + 0.028(60)}}{1 + e^{-3.059 + 0.028(60)}}$$

$$= 0.1969$$

The endpoints of the confidence interval for the estimated logistic probability can be obtained by following a similar process using the endpoints of the estimated logit. The lower limit is

$$\frac{e^{-1.7741}}{1 + e^{-1.7741}} = 0.1450$$

And the upper limit is:

$$\frac{e^{-1.0378}}{1 + e^{-1.0378}} = 0.2616$$

The estimated logistic probability of dying in the ICU for a 60 year old subject, 0.1969, is an estimate of the proportion of 60 year old subjects in the population sampled that die in the ICU. The confidence interval suggests that this mean could be as low as 0.1450 or as high as 0.2616 with 95% confidence.

(j)   *Use the logistic regression package to obtain the estimated logit and its standard error for each subject in the ICU study. Graph the estimated logit and the pointwise 95 percent confidence limits versus AGE for each subject. Explain (in words), the similarities and differences between the appearance of this graph and a graph of a fitted linear regression model and its pointwise 95 percent confidence bands.*

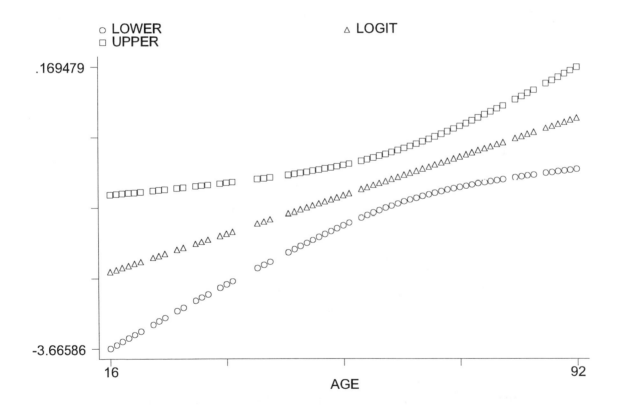

In both the above graph and in a graph of a fitted linear regression model and its pointwise 95% confidence bands, the confidence interval bands are hyperbolic in shape. In linear regression, the minimum width confidence interval is always found at the mean value of the independent variable. In logistic regression, this is not the case.

2. *Use the ICU Study and repeat Exercises 1(a), 1(b), 1(d), 1(e) and 1(g) using the variable "type of admission," TYP, as the covariate.*

   (a) *Write down the equation for the logistic regression model of STA on TYP. Write down the equation for the logit transformation of this logistic regression model. What characteristic of the outcome variable, STA, leads us to consider the logistic regression model as opposed to the usual linear regression model to describe the relationship between STA and TYP?*

   Logistic regression model:

   $$\pi(TYP) = \frac{e^{\beta_0 + \beta_1 *(TYP)}}{1 + e^{\beta_0 + \beta_1 *(TYP)}}$$

   Logit transformation:

   $$g(TYP) = \beta_0 + \beta_1 * (TYP)$$

   We consider the logistic regression model, rather than the linear regression model to describe the relationship between STA and TYP because the outcome variable, STA, is dichotomous, taking on the values 0 and 1.

   (b) *Form a scatterplot of STA versus TYP.*

. graph sta typ

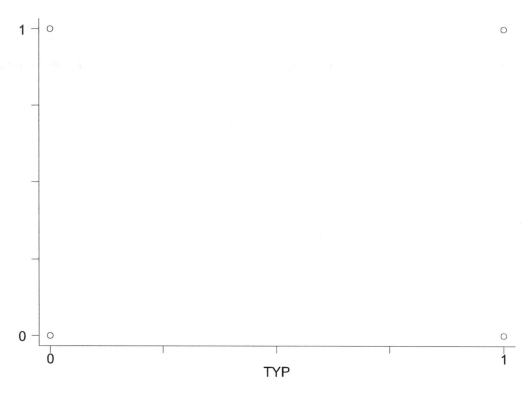

```
. tab STA TYP

          |       TYP
    STA  |        0           1  |      Total
---------+----------------------+----------
      0  |       51         109  |        160
      1  |        2          38  |         40
---------+----------------------+----------
   Total |       53         147  |        200
```

(c)   Write down an expression for the likelihood and log likelihood for the logistic regression model in Exercise 2(a) using the ungrouped, n = 200, data. Obtain expression for the two likelihood equations.

likelihood function:

$$l(\beta) = \prod_{i=1}^{n} \zeta(x_i) \qquad \text{where, } \zeta(x_i) = \pi(x_i)^{y_i} \left[1 - \pi(x_i)\right]^{1-y_i}$$

$$x = \text{TYP} \quad \text{and} \quad y_i = \begin{cases} 0 & \text{if the patient lived} \\ 1 & \text{if the patient died} \end{cases}$$

log likelihood function:

$$L(\beta) = \ln[l(\beta)] = \sum_{i=1}^{n} \{ y_i \ln[\pi(x_i)] + (1 - y_i)\ln[1 - \pi(x_i)] \}$$

likelihood equations:

$$\sum_{i=1}^{n} [y_i - \pi(x_i)] = 0$$

and

$$\sum_{i=1}^{n} x_i [y_i - \pi(x_i)] = 0$$

(d)   *Using a logistic regression package of your choice, obtain the maximum likelihood estimates of the parameters of the logistic regression model in Exercise 2(a). These estimates should be based on the ungrouped, n=200, data. Using these estimates, write down the equation for the fitted values, that is, the estimated logistic probabilities. Plot the equation for the fitted values on the axes used in the scatterplots in Exercise 2(b).*

```
. logit STA TYP

Iteration 0:   Log Likelihood =-100.08048
Iteration 1:   Log Likelihood =-93.425171
Iteration 2:   Log Likelihood =-92.585562
Iteration 3:   Log Likelihood =-92.525071
Iteration 4:   Log Likelihood =-92.524467
Iteration 5:   Log Likelihood =-92.524467

Logit Estimates                                Number of obs =    200
                                               chi2(1)       =  15.11
                                               Prob > chi2   = 0.0001
Log Likelihood = -92.524467                    Pseudo R2     = 0.0755


---------------------------------------------------------------------
    STA |      Coef.   Std. Err.       z     P>|z|    [95% Conf. Interval]
--------+------------------------------------------------------------
    TYP |   2.184917   .7450489     2.933    0.003     .7246476   3.645186
  _cons |  -3.238678   .7208383    -4.493    0.000    -4.651496  -1.825861
---------------------------------------------------------------------
```

logistic regression model (fitted values):

$$\pi(TYP) = \frac{e^{-3.238678+2.184917*(TYP)}}{1+e^{-3.238678+2.184917*(TYP)}}$$

logit transformation:

$$g(TYP) = -3.238678 + 2.184917 * (TYP)$$

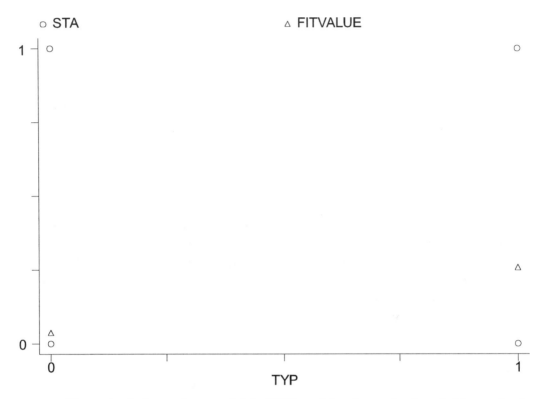

Since the independent variable TYP and the dependent variable are both dichotomous, there are only four possible combinations of these variables. This is indicated clearly in the scatterplot from Exercise 2(b). The frequency of each combination is indicated in the 2x2 contingency table from Exercise 2(b) which shows that 147 people have TYP=1 (emergency admissions). The most common combination is TYP=1, STA=0 which describes persons who were admitted to the ICU as an emergency and who were discharged alive.

The above scatterplot indicates that the probability of being discharged dead is greater among those patients whose admission to the ICU was on an emergency basis.

(e)   *Using the results of the output from the logistic regression package used for Exercise 2(d), assess the significance of the slope coefficient for TYP using the likelihood ratio test, the Wald test, and, if possible, the Score test. What assumptions are needed for the p-values computed from each of these tests to be valid? Are the results of these tests consistent with one another? What is the value of the deviance of the fitted model?*

Deviance:

$$D = -2\ln\left[\frac{\text{(likelihood of the current model)}}{\text{(likelihood of the saturated model)}}\right]$$

$$D = -2(-92.524467)$$

$$D = 185.048934$$

Likelihood Ratio Test:

$$H_0: \beta_1 = 0$$

$$H_A: \beta_1 \neq 0$$

$$G = D(\text{model without variable}) - D(\text{model with variable})$$

$$G = 200.16 - 185.048934$$

$$G = 15.11 \qquad\qquad G \sim \chi^2(1) \qquad\qquad p \leq 0.0001$$

$\therefore$ reject H_0, it is not consistent with the data that $\beta_1 = 0$; we conclude that TYP is a significant predictor of STA.

Assumption: the statistic G will follow a χ^2 distribution with 1 degree of freedom under the null hypothesis.

Wald Test:

$$H_0: \beta_1 = 0$$

$$H_A: \beta_1 \neq 0$$

$$W = \frac{\hat{\beta_1}}{S\hat{E}(\hat{\beta_1})} = 2.933 \quad W \sim N(0,1) \qquad\qquad p = 0.003$$

$\therefore$ reject H_0, it is not consistent with the data that $\beta_1 = 0$; we conclude that AGE is a significant predictor of STA.

Assumption: the Wald statistic will follow a normal distribution with mean 0 and variance 1, when n is large

Score Test:

$$H_0 : \beta_1 = 0$$

$$H_A : \beta_1 \neq 0$$

$$ST = \frac{\sum_{i=1}^{n} x_i (y_i - \bar{y})}{\sqrt{\bar{y}(1 - \bar{y}) \sum_{i=1}^{n} (x_i - \bar{x})^2}}$$

$$ST = \frac{8.6}{\sqrt{0.16(38.955)}} = \frac{8.6}{2.497} = 3.44 \quad p = 0.00058$$

$$ST \sim N(0,1)$$

$\therefore$ reject H_0, it is not consistent with the data that $\beta_1 = 0$; we conclude that AGE is a significant predictor of STA.

Assumption: The score statistic is normally distributed with mean 0 and variance 1.

The results of the likelihood ratio test, the Wald test and the Score test are consistent. Each test indicates that the model is significant.

The value of the deviance for the fitted model is $D = 185.05$.

3. In the Low Birth Weight Study described in Section 1.6.2, one variable that physicians felt was important to control for was the weight of the mother at the last menstrual period, LWT. Repeat steps (a)-(g) of Exercise 1, but for Exercise 3(c) use intervals [80,99], [100,109], [110, 114], [115, 119], [120, 124], [125, 129], [130, 250].

    (a)   *Write down the equation for the logistic regression model of LOW on LWT. Write down the equation for the logit transformation of this logistic regression model. What characteristic of the outcome variable, LOW, leads us to consider the logistic regression model as opposed to the usual linear regression model to describe the relationship between LOW and LWT?*

Logistic regression model:

$$\pi(LWT) = \frac{e^{\beta_0 + \beta_1 * (LWT)}}{1 + e^{\beta_0 + \beta_1 * (LWT)}}$$

Logit transformation:

$$g(LWT) = \beta_0 + \beta_1 * (LWT)$$

We consider the logistic regression model, rather than the usual linear regression model to describe the relationship between LOW and LWT because the outcome variable, LOW, is dichotomous, taking on the values 0 and 1.

(b)   *Form a scatterplot of LOW versus LWT.*

. `graph low lwt`

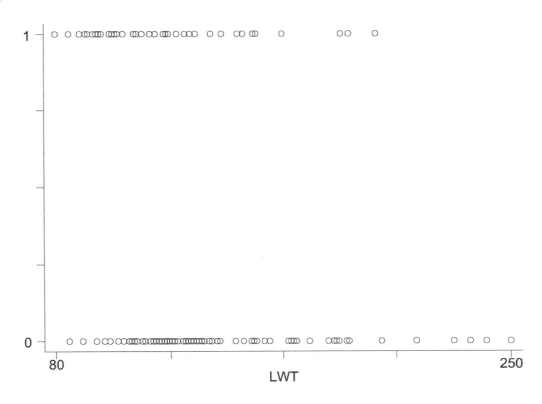

(c)   *Using the intervals [80,99], [100,109], [110, 114], [115, 119], [120, 124], [125, 129], [130, 250] for LWT, compute the LOW mean over subjects within each LWT interval. Plot these values of mean LOW versus the midpoint of the LWT interval using the same set of axes as was used in Exercise 3(b).*

LWTGP	Interval	Midpoint	n	Mean LOW
1	80-99	90.0	19	0.474
2	100-109	105.0	23	0.522
3	110-114	112.5	18	0.278
4	115-119	117.5	15	0.200
5	120-124	122.5	28	0.250
6	125-129	127.5	7	0.286
7	130-250	190.5	79	0.266

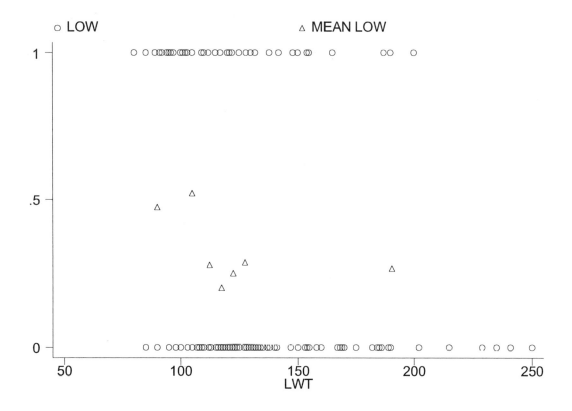

(d)   *Write down an expression for the likelihood and log likelihood for the logistic regression model in Exercise 3(a) using the ungrouped, n=189, data. Obtain expression for the two likelihood equations.*

likelihood function:

$$l(\beta) = \prod_{i=1}^{n} \zeta(x_i) \qquad \text{where, } \zeta(x_i) = \pi(x_i)^{y_i} \left[1 - \pi(x_i)\right]^{1-y_i}$$

$$x = \text{LWT} \qquad \text{and} \qquad y_i = \begin{cases} 0 & \text{normal birth weight} \\ 1 & \text{low birth weight} \end{cases}$$

log likelihood function:

$$L(\beta) = \ln[l(\beta)] = \sum_{i=1}^{n} \left\{ y_i \ln[\pi(x_i)] + (1 - y_i)\ln[1 - \pi(x_i)] \right\}$$

likelihood equations:

$$\sum_{i=1}^{n} [y_i - \pi(x_i)] = 0$$

and

$$\sum_{i=1}^{n} x_i [y_i - \pi(x_i)] = 0$$

(e)   *Using a logistic regression package of your choice, obtain the maximum likelihood estimates of the parameters of the logistic regression model in Exercise 3(a). These estimates should be based on the ungrouped, n=189, data. Using these estimates, write down the equation for the fitted values, that is, the estimated logistic probabilities. Plot the equation for the fitted values on the axes used in the scatterplots in Exercises 3(b) and 3(c).*

```
. logit low lwt

Iteration 0:   Log Likelihood =   -117.336
Iteration 1:   Log Likelihood =-114.41626
Iteration 2:   Log Likelihood =-114.34546
Iteration 3:   Log Likelihood =-114.34533

Logit Estimates                              Number of obs =      189
                                             chi2(1)       =     5.98
                                             Prob > chi2   = 0.0145
Log Likelihood = -114.34533                  Pseudo R2     = 0.0255

------------------------------------------------------------------------------
    low |     Coef.   Std. Err.       z     P>|z|     [95% Conf. Interval]
--------+---------------------------------------------------------------------
    lwt | -.0140583   .0061696    -2.279   0.023    -.0261504   -.0019661
  _cons |  .9983143   .7852889     1.271   0.204    -.5408235    2.537452
------------------------------------------------------------------------------
```

logistic regression model (fitted values):

$$\pi(LWT) = \frac{e^{0.9983143+(-0.0140583)*(LWT)}}{1+e^{0.9983143+(-0.0140583)*(LWT)}}$$

logit transformation:

$$g(LWT) = 0.9983143 + (-0.0140583)*(LWT)$$

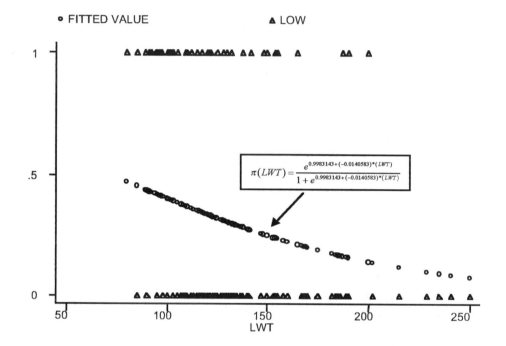

(f)   *Summarize (describe in words) the results presented in the plot obtained from Exercises 3(b), 3(c) and 3(e).*

The plot of LOW vs. LWT indicated in the scatterplot 3(b) demonstrates the dichotomous nature of the LOW variable, which takes on the value zero when the birthweight of the baby was greater than or equal to 2500 g and the value one when the birthweight was less than 2500 g (low birthweight). The plot suggests that women who were lighter at their last menstrual periods were more likely to give birth to low birthweight babies than heavier women, although overall, women are more likely to give birth to normal weight babies.

The plot of mean LOW by weight at last menstrual period group (indicated in the scatterplot in 3(c) by triangles) reinforces this impression. In general, as the mother's weight at last menstrual period increases, the probability of giving birth to a low birthweight baby decreases.

The plot of the estimated logistic probabilities vs. LWT (indicated in the scatterplot 3(d) by circles) indicates that the probability of giving birth to a low birthweight baby does decrease with increasing maternal weight.

(g)   *Using the results of the output from the logistic regression package used for Exercise 3(e), assess the significance of the slope coefficient for LWT using the likelihood ratio test, the Wald test, and, if possible the Score test. What assumptions are needed for the p-values computed for each of these tests to be valid? Are the results of these tests consistent with one another? What is the value of the deviance for the fitted model?*

Deviance:

$$D = -2 \ln \left[\frac{(\text{likelihood of the current model})}{(\text{likelihood of the saturated model})} \right]$$

$$D = -2(-114.34533)$$

$$D = 228.69066$$

Likelihood Ratio Test:

$H_0 : \beta_1 = 0$

$H_A : \beta_1 \neq 0$

$G = D(\text{model without variable}) - D(\text{model with variable})$

$G = 234.672 - 228.69066$

$G = 5.98 \qquad\qquad G \sim \chi^2(1) \qquad\qquad p = 0.0145$

$\therefore$ reject H_0, it is not consistent with the data that $\beta_1 = 0$; we conclude that LWT is a significant predictor of LOW.

Assumption: the statistic G will follow a χ^2 distribution with 1 degree of freedom under the null hypothesis.

Wald Test:

$H_0 : \beta_1 = 0$

$H_A : \beta_1 \neq 0$

$$W = \frac{\hat{\beta_1}}{\hat{SE}(\hat{\beta_1})} = -2.279 \qquad W \sim N(0,1) \qquad p = 0.023$$

$\therefore$ reject H$_0$, it is not consistent with the data that $\beta_1 = 0$; we conclude that LWT is a significant predictor of LOW.

Assumption: the Wald statistic will follow a normal distribution with mean 0 and variance 1.

Score Test:

$$H_0 : \beta_1 = 0$$

$$H_A : \beta_1 \neq 0$$

$$ST = \frac{\sum_{i=1}^{n} x_i (y_i - \bar{y})}{\sqrt{\bar{y}(1 - \bar{y}) \sum_{i=1}^{n} (x_i - \bar{x})^2}}$$

$$ST = \frac{-453.1}{\sqrt{0.21(175798.5)}} = \frac{-453.1}{194.3} = -2.332 \quad p = 0.010$$

$$ST \sim N(0,1)$$

$\therefore$ reject H$_0$, it is not consistent with the data that $\beta_1 = 0$; we conclude that LWT is a significant predictor of LOW.

Assumption: The score statistic is normally distributed with mean 0 and variance 1.

The results of the likelihood ratio test, the Wald test and the Score test are consistent. Each test indicates that the model is significant.

The value of the deviance for the fitted model is $D = 228.69$.

(h)   *The graph in 3(c) does not look "S-Shaped". The primary reason is that the range of plotted values is from approximately 0.2 to 0.56. Explain why a model for the probability of low birth weight as a function of LWT could still be the logistic regression model.*

The dichotomous nature of the variable LOW is demonstrated in the graph. LOW is 0 if birth weight is greater than 2500 grams and 1 otherwise. The plot suggests that mothers weighing between 100 and 150 lbs are more likely to have a low birth weight child. Those women weighing > 150 lbs do not have as high a risk of low birth weight babies. The scatterplot reinforces this statement. The S-shaped form of the model is not evident.

4. *In the Prostate Cancer Study described in Section 1.6.3, one variable thought to be particularly predictive of capsule penetration is the prostate specific antigen level, PSA. Repeat steps (a)-(g) and (j) of Exercise 1, using CAPSULE as the outcome variable and PSA as the covariate. For Exercise 4(c) use intervals for PSA of [0, 2.4], [2.5, 4.4], [4.5, 6.4], [6.5, 8.4], [8.5, 10.4], [10.5, 12.4], [12.5, 20.4], [20.5, 140].*

    (a)   *Write down the equation for the logistic regression model of CAPSULE on PSA. Write down the equation for the logit transformation of this logistic regression model. What characteristic of the outcome variable, CAPSULE, leads us to consider the logistic regression model as opposed to the usual linear regression model to describe the relationship between CAPSULE and PSA?*

       Logistic regression model:

$$\pi(PSA) = \frac{e^{\beta_0 + \beta_1 *(PSA)}}{1 + e^{\beta_0 + \beta_1 *(PSA)}}$$

       Logit transformation:

$$g(PSA) = \beta_0 + \beta_1 * (PSA)$$

       We consider the logistic regression model, rather than the usual linear regression model to describe the relationship between CAPSULE and PSA because the outcome variable, CAPSULE, is dichotomous, taking on the values 0 and 1.

    (b)   *Form a scatterplot of CAPSULE versus PSA.*

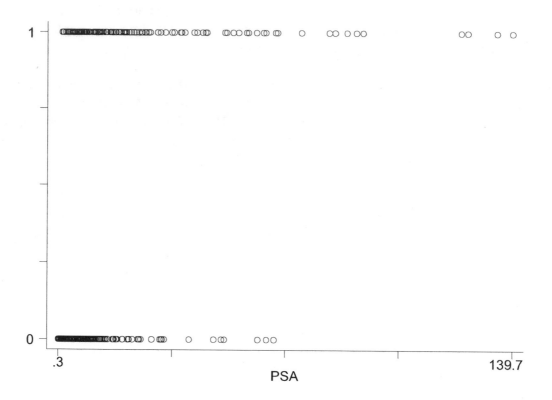

(c)   *Using the intervals [0,2.4], [2.5, 4.4], [4.5, 6.4], [6.5, 8.4], [8.5, 10.4], [10.5, 12.4], [12.5, 20.4], [20.5, 140] for PSA, compute the CAPSULE mean over subjects within each PSA interval. Plot these values of mean CAPSULE versus the midpoint of the PSA interval using the same set of axes as was used in Exercise 4(b).*

PSAGP	Interval	Midpoint	n	Mean Capsule
1	0-2.4	1.25	33	0.091
2	2.5-4.4	3.5	39	0.308
3	4.5-6.4	5.5	57	0.246
4	6.5-8.4	7.5	54	0.444
5	8.5-10.4	9.5	39	0.308
6	10.5-12.4	11.5	28	0.214
7	12.5-20.4	16.5	56	0.554
8	20.5-140	80.5	74	0.689

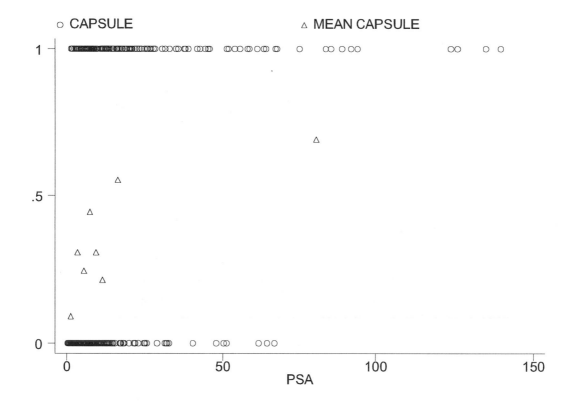

○ CAPSULE   △ MEAN CAPSULE

(d)   *Write down an expression for the likelihood and log likelihood for the logistic regression model in Exercise 1(a) using the ungrouped, n=380, data. Obtain expression for the two likelihood equations.*

likelihood function:

$$l(\beta) = \prod_{i=1}^{n} \zeta(x_i) \qquad \text{where, } \zeta(x_i) = \pi(x_i)^{y_i}\left[1 - \pi(x_i)\right]^{1-y_i}$$

$$x = \text{PSA} \qquad \text{and} \qquad y_i = \begin{cases} 0 & \text{if no penetration} \\ 1 & \text{if penetration} \end{cases}$$

log likelihood function:

$$L(\beta) = \ln[l(\beta)] = \sum_{i=1}^{n}\left\{ y_i \ln[\pi(x_i)] + (1 - y_i)\ln[1 - \pi(x_i)]\right\}$$

likelihood equations:

$$\sum_{i=1}^{n}\left[y_i - \pi(x_i)\right] = 0$$

and

$$\sum_{i=1}^{n} x_i\left[y_i - \pi(x_i)\right] = 0$$

(e)   *Using a logistic regression package of your choice, obtain the maximum likelihood estimates of the parameters of the logistic regression model in Exercise 4(a). These estimates should be based on the ungrouped, n=380, data. Using these estimates, write down the equation for the fitted values, that is, the estimated logistic probabilities. Plot the equation for the fitted values on the axes used in the scatterplots in Exercises 4(b) and 4(c).*

```
. logit capsule psa

Iteration 0:   Log Likelihood =-256.14442
Iteration 1:   Log Likelihood =-233.39251
Iteration 2:   Log Likelihood =-231.64579
Iteration 3:   Log Likelihood =-231.58066
Iteration 4:   Log Likelihood =-231.58055

Logit Estimates                                Number of obs =      380
                                               chi2(1)       =    49.13
                                               Prob > chi2   = 0.0000
Log Likelihood = -231.58055                    Pseudo R2     = 0.0959


------------------------------------------------------------------------------
 capsule |     Coef.   Std. Err.       z     P>|z|     [95% Conf. Interval]
---------+--------------------------------------------------------------------
     psa |   .0501761   .0092502     5.424   0.000     .032046     .0683062
   _cons |  -1.113695   .1615629    -6.893   0.000    -1.430352   -.7970371
------------------------------------------------------------------------------
```

logistic regression model (fitted values):

$$\hat{\pi}(PSA) = \frac{e^{-1.113695+(0.0501761)*(PSA)}}{1 + e^{-1.113695+(0.0501761)*(PSA)}}$$

logit transformation:

$$g(PSA) = -1.113695 + (0.0501761)*(PSA)$$

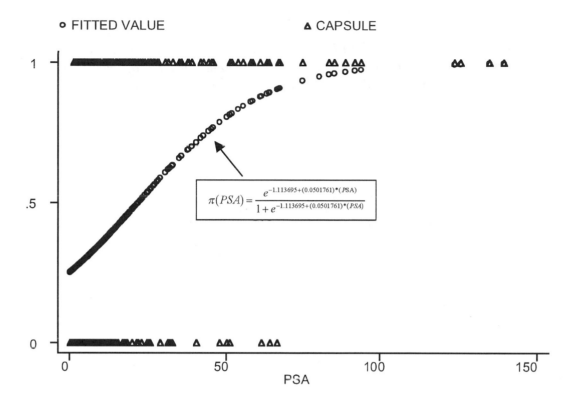

(f)   *Summarize (describe in words) the results presented in the plot obtained from Exercises 4(b), 4(c) and 4(e).*

The plot of CAPSULE vs. PSA (indicated in the scatterplot in 4(b) demonstrates the dichotomous nature of the CAPSULE variable, which takes on the value zero when there is no tumor penetration of the prostatic capsule and the value one when there is tumor penetration of the prostatic capsule. The plot suggests that men with lower values of PSA are more likely not to have tumor penetration of the prostatic capsule than men with higher PSA values.

The plot of mean CAPSULE by PSA group (indicated in the scatterplot in 4(c) by triangles) reinforces this impression. In general, as the PSA value increases, the probability of having tumor penetration of the prostatic capsule increases.

The plot of the estimated logistic probabilities vs. PSA (indicated in the scatterplot in 4(d) by circles) indicates that the probability of tumor penetration of the prostatic capsule does increase with increasing PSA.

(g)   *Using the results of the output from the logistic regression package used for Exercise 4(e), assess the significance of the slope coefficient for PSA using the likelihood ratio test, the Wald test, and, if possible the Score test. What assumptions are needed for the p-values computed for each of these tests to be valid? Are the results of these tests consistent with one another? What is the value of the deviance for the fitted model?*

Deviance:

$$D = -2 \ln \left[\frac{\text{(likelihood of the current model)}}{\text{(likelihood of the saturated model)}} \right]$$

$$D = -2(-231.58055)$$

$$D = 463.1611$$

Likelihood Ratio Test:

$$H_0 : \beta_1 = 0$$

$$H_A : \beta_1 \neq 0$$

$$G = D(\text{model without variable}) - D(\text{model with variable})$$

$$G = 512.28884 - 463.1611$$

$$G = 49.128 \qquad G \sim \chi^2(1) \qquad p < 0.0001$$

$\therefore$ reject H_0, it is not consistent with the data that $\beta_1 = 0$; we conclude that PSA is a significant predictor of CAPSULE.

Assumption: the statistic G will follow a χ^2 distribution with 1 degree of freedom under the null hypothesis.

Wald Test:

$$H_0 : \beta_1 = 0$$

$$H_A : \beta_1 \neq 0$$

$$W = \frac{\hat{\beta_1}}{S\hat{E}(\hat{\beta_1})} = 5.424 \quad W \sim N(0,1) \qquad p < 0.0001$$

$\therefore$ reject H_0, it is not consistent with the data that $\beta_1 = 0$; we conclude that PSA is a significant predictor of CAPSULE.

Assumption: the Wald statistic will follow a normal distribution with mean 0 and variance 1.

Score Test:

$$H_0 : \beta_1 = 0$$

$$H_A : \beta_1 \neq 0$$

$$ST = \frac{\sum_{i=1}^{n} x_i (y_i - \bar{y})}{\sqrt{\bar{y}(1 - \bar{y}) \sum_{i=1}^{n} (x_i - \bar{x})^2}}$$

$$ST = \frac{1233.569}{\sqrt{0.24(151563.2)}} = \frac{1233.569}{190.9} = 6.460879 \qquad p < 0.0001$$

$$ST \sim N(0,1)$$

$\therefore$ reject H_0, it is not consistent with the data that $\beta_1 = 0$; we conclude that PSA is a significant predictor of CAPSULE.

Assumption: The score statistic is normally distributed with mean 0 and variance 1.

The results of the likelihood ratio test, the Wald test and the Score test are consistent. Each test indicates that the model is significant.

The value of the deviance for the fitted model is $D = 463.16$.

(j)   *Use the logistic regression package to obtain the estimated logit and its standard error for each subject in the Prostate Cancer Study. Graph the estimated logit and the pointwise 95 percent confidence limits versus PSA for each subject. Explain (in words) the similarities and differences between the appearance of this graph and a graph of a fitted linear regression model and its pointwise 95% confidence bands.*

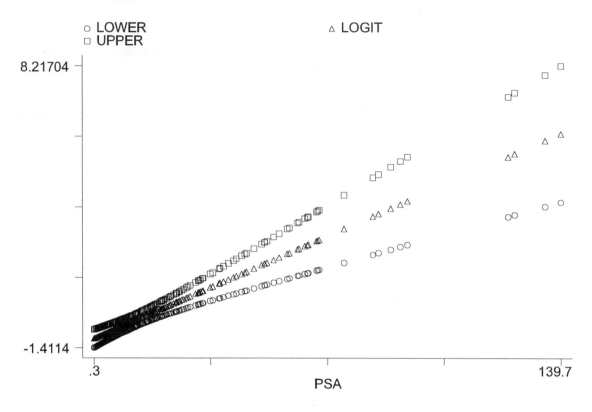

In both the above graph and in a graph of a fitted linear regression model and its pointwise 95% confidence bands, the confidence interval bands are hyperbolic in shape. In linear regression, the minimum width confidence interval is always found at the mean value of the independent variable. In logistic regression, this is not the case.

Chapter Two – Solutions

1. Use the ICU data described in Section 1.6.1 and consider the multiple logistic regression model of vital status, STA, on age (AGE), cancer part of the present problem (CAN), CPR prior to ICU admission (CPR), infection probable at ICU admission (INF), and race (RACE).

    (a)   The variable RACE is coded at three levels. Prepare a table showing the coding of the two design variables necessary for including this variable in a logistic regression model.

RACE	Label	RACE_2	RACE_3
1	White	0	0
2	Black	1	0
3	Other	0	1

    (b)   Write down the equation for the logistic regression model of STA on AGE, CAN, CPR, INF and RACE. Write down the equation for the logit transformation of this logistic regression model. How many parameters does this model contain?

    logistic regression model:

    $$\pi(\mathbf{x}) = \frac{e^{\beta_0 + \beta_1*(AGE) + \beta_2*(CAN) + \beta_3*(CPR) + \beta_4*(INF) + \beta_5*(RACE\_2) + \beta_6*(RACE\_3)}}{1 + e^{\beta_0 + \beta_1*(AGE) + \beta_2*(CAN) + \beta_3*(CPR) + \beta_4*(INF) + \beta_5*(RACE\_2) + \beta_6*(RACE\_3)}}$$

    where $\mathbf{x}$ = vector of covariates

    logit transformation:

    $$g(\mathbf{x}) = \beta_0 + \beta_1*(AGE) + \beta_2*(CAN) + \beta_3*(CPR) + \beta_4*(INF) + \beta_5*(RACE\_2) + \beta_6*(RACE\_3)$$

    This model contains 7 parameters.

    (c)   Write down an expression for the likelihood and log likelihood for the logistic regression model in Exercise 1(b). How many likelihood equations are there? Write down an expression for a typical likelihood equation for this problem.

    likelihood function:

    $$l(\beta) = \prod_{i=1}^{n} \zeta(\mathbf{x}_i) \qquad \text{where,} \quad \zeta(\mathbf{x}_i) = \pi(\mathbf{x}_i)^{y_i} \left[1 - \pi(\mathbf{x}_i)\right]^{1 - y_i}$$

    $\mathbf{x}$ = set of covariates   and $y_i = \begin{cases} 0 & \text{if the patient lived} \\ 1 & \text{if the patient died} \end{cases}$

log likelihood function:

$$L(\beta) = \ln\left[l(\beta)\right] = \sum_{i=1}^{n}\left\{y_i \ln\left[\pi(\mathbf{x}_i)\right] + \left(1 - y_i\right)\ln\left[1 - \pi(\mathbf{x}_i)\right]\right\}$$

There will be $p+1$ or 7 likelihood equations for this problem.

Likelihood equations that result may be expressed as follows:

$$\sum_{i=1}^{n}\left[y_i - \pi(\mathbf{x}_i)\right] = 0$$

and

$$\sum_{i=1}^{n} x_i\left[y_i - \pi(\mathbf{x}_i)\right] = 0$$

(d)   *Using a logistic regression package, obtain the maximum likelihood estimates of the parameters of the logistic regression model in Exercise 1(b). Using these estimates, write down the equation for the fitted values, that is the estimated logistic probabilities.*

```
. logit sta age can cpr inf r_2 r_3

Iteration 0:   Log Likelihood =-100.08048
Iteration 1:   Log Likelihood =-90.619912
Iteration 2:   Log Likelihood =-89.663593
Iteration 3:   Log Likelihood =-89.650384
Iteration 4:   Log Likelihood =-89.650364

Logit Estimates                                Number of obs =      200
                                               chi2(6)       =    20.86
                                               Prob > chi2   =   0.0019
Log Likelihood = -89.650364                    Pseudo R2     =   0.1042


------------------------------------------------------------------------------
    sta  |     Coef.    Std. Err.       z      P>|z|     [95% Conf. Interval]
---------+--------------------------------------------------------------------
    age  |   .0271207    .0115879     2.340    0.019     .0044089    .0498325
    can  |   .2445106    .616815      0.396    0.692    -.9644246   1.453446
    cpr  |   1.646497    .6234135     2.641    0.008     .424629    2.868365
    inf  |   .6806676    .3804176     1.789    0.074    -.0649372   1.426272
    r_2  |  -.9570777   1.084467     -0.883    0.377    -3.082593   1.168438
    r_3  |   .2597493    .8712682     0.298    0.766    -1.447905   1.967404
   _cons |  -3.51152     .8144295    -4.312    0.000    -5.107772  -1.915267
------------------------------------------------------------------------------

. lrtest, saving(0)
```

logit transformation:

$$g(x) = -3.512 + 0.027 * (AGE) + 0.245 * (CAN) + 1.646(CPR)$$

$$+0.681(INF) - 0.957(RACE\_2) + 0.260(RACE\_3)$$

logistic regression model:

$$\pi(x) = \frac{e^{-3.512+0.027*(AGE)+0.245*(CAN)+1.646(CPR)+0.681(INF)-0.957*(RACE\_2)+0.260(RACE\_3)}}{1+e^{-3.512+0.027*(AGE)+0.245*(CAN)+1.646(CPR)+0.681(INF)-0.957*(RACE\_2)+0.260(RACE\_3)}}$$

(e)   Using the results of the output from the logistic regression package used in Exercise 1(d), assess the significance of the slope coefficients for the variables in the model using the likelihood ratio test. What assumptions are needed for the p-values computed for this test to be valid? What is the value of the deviance for the fitted model?

Deviance:

$$D = -2\ln\left[\frac{(\text{likelihood of the current model})}{(\text{likelihood of the saturated model})}\right]$$

$$D = -2(-89.650364)$$

$$D = 179.300728$$

Likelihood Ratio Test:

$$H_0 : \beta_1 = \beta_2 = \beta_3 = \beta_4 = \beta_5 = \beta_6 = 0$$

$$H_A : \text{At least one coefficient is not equal to 0}$$

$$G = D(\text{model without variable}) - D(\text{model with variable})$$

$$G = 200.16 - 179.300728$$

$$G = 20.86 \qquad\qquad G \sim \chi^2(6) \qquad\qquad p = 0.00194$$

∴ reject H_0, it is not consistent with the data that all β=0; we conclude that together, AGE, CAN, CPR, INF, and RACE are significant predictors of STA.

Assumption: the statistic G will follow a χ^2 distribution with 6 degrees of freedom under the null hypothesis.

The value of the deviance for the fitted model is: $D = 179.30$.

(f) *Use the Wald statistics to obtain an approximation to the significance of the individual slope coefficients for the variables in the model. Fit a reduced model that eliminates those variables with nonsignificant Wald statistics. Assess the joint (conditional) significance of the variables excluded from the model. Present the results of fitting the reduced model in a table.*

Using a critical value of $p < 0.05$, based on the computer output from Exercise 1(d), one can conclude that the variables AGE, CPR and possibly INF are significant while CAN, RACE_2 and RACE_3 are not significant. A reduced model was fit containing only those variables thought to be significant:

```
. logit sta age cpr inf

Iteration 0:   Log Likelihood =-100.08048
Iteration 1:   Log Likelihood =-91.107974
Iteration 2:   Log Likelihood =-90.262687
Iteration 3:   Log Likelihood =-90.256715
Iteration 4:   Log Likelihood =-90.256714

Logit Estimates                              Number of obs =      200
                                             chi2(3)       =    19.65
                                             Prob > chi2   = 0.0002
Log Likelihood = -90.256714                  Pseudo R2     = 0.0982


------------------------------------------------------------------------
    sta |      Coef.   Std. Err.       z     P>|z|     [95% Conf. Interval]
--------+---------------------------------------------------------------
    age |    .027922   .0113598    2.458    0.014     .0056573    .0501867
    cpr |   1.630662   .6155313    2.649    0.008     .4242431    2.837081
    inf |   .6970764      .3775    1.847    0.065    -.0428101    1.436963
  _cons |  -3.576045   .7730606   -4.626    0.000    -5.091216   -2.060874

. lrtest
Logit:  likelihood-ratio test                chi2(3)       =      1.21
                                             Prob > chi2   =    0.7500
```

The likelihood ratio test comparing the above model with the full model will have a distribution that is chi-square with 3 degrees of freedom under the hypothesis that the coefficients for the variables that were excluded are equal to zero.

Likelihood Ratio Test:

H_0: Coefficients for eliminated variables all equal 0

H_A: At least one coefficient is not equal to 0

$G = D(\text{model without variables}) - D(\text{model with variables})$

$$G = 180.513428 - 179.300728$$

$$G = 1.21 \qquad G \sim \chi^2(3) \qquad p = 0.75061$$

$\therefore$ do not reject H_0, it is consistent with the data that the coefficients for the eliminated variables are all equal to zero; we conclude that the reduced model containing only AGE, CPR and INF is as good as the full model. Statistically speaking, there is no advantage to including CAN or RACE in the model.

Based on a strict definition of significance at p<0.05, one would also exclude the variable INF from the model. The reduced model contains AGE and CPR only.

```
. logit sta age cpr

Iteration 0:   Log Likelihood =-100.08048
Iteration 1:   Log Likelihood =-92.714062
Iteration 2:   Log Likelihood = -91.98047
Iteration 3:   Log Likelihood = -91.97634
Iteration 4:   Log Likelihood = -91.97634

Logit Estimates                           Number of obs =      200
                                          chi2(2)       =    16.21
                                          Prob > chi2   =   0.0003
Log Likelihood =   -91.97634              Pseudo R2     =   0.0810

---------------------------------------------------------------------
    sta |      Coef.   Std. Err.       z    P>|z|    [95% Conf. Interval]
--------+------------------------------------------------------------
    age |   .0296074   .0111489    2.656   0.008    .0077559    .0514589
    cpr |   1.784092   .6072971    2.938   0.003    .5938116    2.974373
  _cons |  -3.351956   .7454995   -4.496   0.000   -4.813108   -1.890803
---------------------------------------------------------------------
```

The likelihood ratio test comparing the above model with the model that included INF will have a distribution that is chi-square with 1 degree of freedom under the hypothesis that the coefficient for INF is equal to zero.

Likelihood Ratio Test:

H_0: Coefficient for INF equals 0

H_A: Coefficient for INF is not equal to 0

$G = D(\text{model without variable}) - D(\text{model with variable})$

$G = 183.95 - 180.51$

$$G = 3.44 \qquad G \sim \chi^2(1) \qquad p = 0.06364$$

∴ do not reject H_0, it is consistent with the data that the coefficient for the eliminated variable, INF, is equal to zero; we conclude that the reduced model containing only AGE and CPR is as good as the full model. Statistically speaking, there is no advantage to including INF in the model. However, the significance of INF is borderline. The decision to include INF should be made after considering clinical reasons for its inclusion in the model.

Estimated Coefficients for a Multiple Logistic Regression Model for STA using the Variables AGE and CPR from the ICU Study (*n*=200)

```
. logit STA AGE CPR

Iteration 0:    log likelihood = -100.08048
Iteration 1:    log likelihood = -92.714062
Iteration 2:    log likelihood =  -91.98047
Iteration 3:    log likelihood =  -91.97634
Iteration 4:    log likelihood =  -91.97634

Logit estimates                          Number of obs   =        200
                                         LR chi2(2)      =      16.21
                                         Prob > chi2     =     0.0003
Log likelihood =  -91.97634              Pseudo R2       =     0.0810


------------------------------------------------------------------------------
     STA |      Coef.   Std. Err.       z     P>|z|    [95% Conf. Interval]
---------+--------------------------------------------------------------------
     AGE |   .0296074   .0111489     2.66     0.008    .0077559    .0514589
     CPR |   1.784092   .6072971     2.94     0.003    .5938116    2.974373
   _cons |  -3.351956   .7454995    -4.50     0.000   -4.813108   -1.890803
------------------------------------------------------------------------------
```

(g)   *Using the results from problem 1(f), compute 95 percent confidence intervals for all coefficients in the model. Write a sentence interpreting the confidence intervals for the non-constant covariates.*

Ninety-five percent confidence intervals for the coefficients for AGE and CPR can be computed using equations (1.15) and (1.16) from the text. The intervals shown below are calculated from the STATA output for the logistic regression model shown in problem 1(f). The confidence intervals for the coefficients can also be seen in the 6th and 7th columns presented in this STATA output.

Endpoints of a $100(1-\alpha)\%$ confidence interval for AGE:

$$\hat{\beta}_1 \pm z_{1-\alpha/2} S\hat{E}\left(\hat{\beta}_1\right)$$

$$0.0296 \pm 1.96(0.0111)$$

$$(0.0078, \ 0.0515)$$

Endpoints of a 100(1-α)% confidence interval for CPR:

$$\hat{\beta}_2 \pm z_{1-\alpha/2}S\hat{E}\left(\hat{\beta}_2\right)$$

$$1.7841 \pm 1.96(0.6073)$$

$$(0.5938, \ 2.9744)$$

Endpoints of a 100(1-α)% confidence interval for constant:

$$\hat{\beta}_0 \pm z_{1-\alpha/2}S\hat{E}\left(\hat{\beta}_0\right)$$

$$-3.3520 \pm 1.96(0.7455)$$

$$(-4.8131, \ -1.8908)$$

The 95% confidence interval for AGE suggests that the change in the log odds of dying in the ICU (STA=1) per one year increase in AGE is 0.0296 when the value of CPR is constant and that the change could be as little as 0.0078 or as much as 0.0515 with 95% confidence.

The 95% confidence interval for CPR suggests that the change in the log odds of dying in the ICU (STA=1) for persons who had CPR prior to admission compared with those who had not is 1.7841 when the value of AGE is constant and that the change could be as little as 0.5938 or as much as 2.9744 with 95% confidence.

(h) *Obtain the estimated covariance matrix for the final model fit in Exercise 1(f). Choose a set of values for the covariates in that model and estimate the logit and logistic probability for the population of subjects with these characteristics. Compute 95 percent confidence intervals for the logit and logistic probability. Write a sentence or two interpreting the estimated probability and its confidence interval.*

```
. vce

             |      AGE       CPR     _cons
-------------+------------------------------
         AGE |  .000124
         CPR |  .000811    .36881
       _cons |  -.00802  -.090604    .55577
```

For a patient who is 60 years old and who had CPR prior to ICU admission:

Estimated logit:

$$\hat{g}(x) = \hat{\beta}_0 + \hat{\beta}_1(AGE) + \hat{\beta}_2(CPR)$$

$$= (-3.351956) + (0.0296074) * (60) + (1.784092) * (1)$$

$$= 0.2085812$$

$$Var[\hat{g}(x)] = 0.55577 + (60)^2 * (0.000124) + (1)^2 * (0.36881) + 2 * 60 * (-0.00802)$$
$$+ 2 * 1 * (-0.090604) + 2 * 60 * 1 * (0.000811)$$
$$= 0.3257507$$

$$SE[\hat{g}(x)] = 0.5707457$$

The estimated standard error of the logit for a 60-year old subject with a history of CPR prior to ICU admission, can be used to calculate the endpoints of the confidence interval for the logit for a 60-year old subject with a history of CPR prior to ICU admission:

```
. lincom _b[_cons]+60*_b[AGE]+_b[CPR]*1

 ( 1)   60.0 AGE + CPR + _cons = 0.0

-----------------------------------------------------------------------
       STA |     Coef.   Std. Err.      z    P>|z|    [95% Conf. Interval]
-----------+-----------------------------------------------------------
       (1) |   .2085812   .5707457    0.37   0.715   -.9100598    1.327222
-----------------------------------------------------------------------

. display exp(.2085812)/(1+exp(.2085812))
.55195707

. display exp(-.9100598)/(1+exp(-.9100598))
.2869876

. display exp(1.327222)/(1+exp(1.327222))
.79038075
```

$$\hat{g}(x) \pm z_{1-\alpha/2}\, \hat{SE}[\hat{g}(x)]$$

$$0.2085812 \pm 1.96 * .5707457$$

$$(-.9100598, \ 1.327222)$$

Estimated logistic probability:

$$\hat{\pi}(x) = \frac{e^{\hat{g}(x)}}{1 + e^{\hat{g}(x)}}$$

$$= \frac{e^{0.2085812}}{1 + e^{0.2085812}} = 0.552$$

The endpoints of the confidence interval for the estimated logistic probability can be obtained by following a similar process using the endpoints of the estimated logit.

The lower limit is:

$$\frac{e^{-.9100598}}{1 + e^{-.9100598}} = 0.287$$

And the upper limit is:

$$\frac{e^{1.327222}}{1 + e^{1.327222}} = 0.790$$

The estimated logistic probability of dying in the ICU for a 60 year old subject with history of CPR prior to ICU admission, 0.552, is an estimate of the proportion of 60 year old subjects with history of CPR prior to ICU admission in the population sampled that die in the ICU. The confidence interval suggests that this mean could be as low as 0.287 or as high as 0.790 with 95% confidence.

2.   *Use the Prostate Cancer data described in Section 1.6.3 and consider the multiple logistic regression model of capsule penetration (CAPSULE), on AGE, RACE, results of the digital rectal exam (DPROS and DCAPS), prostate specific antigen (PSA), Gleason score (GLEASON) and tumor volume (VOL).*

(a)   *The variable DPROS is coded at four levels. Prepare a table showing the coding of the three design variables necessary for including this variable in a logistic regression model.*

DPROS	Label	DPROS_2	DPROS_3	DPROS_4
1	No Nodule	0	0	0
2	Unilobar Nodule (Left)	1	0	0
3	Unilobar Nodule (Right)	0	1	0
4	Bilobar Nodule	0	0	1

(b)   *The variable DCAPS is coded 1 and 2. Can this variable be used in its original coding or must a design variable be created? Explore this question by comparing the estimated coefficients obtained from fitting a model containing DCAPS as originally coded and from one using a 0-1 coded design variable, DCAPSnew=DCAPS-1.*

```
. logit capsule age race dpros_2 dpros_3 dpros_4 dcaps psa gleason vol

Iteration 0:   Log Likelihood =-253.29367
Iteration 1:   Log Likelihood =-193.46088
Iteration 2:   Log Likelihood =-187.54897
Iteration 3:   Log Likelihood =-187.26856
Iteration 4:   Log Likelihood =-187.26756

Logit Estimates                              Number of obs =      376
                                             chi2(9)       =   132.05
                                             Prob > chi2   =   0.0000
Log Likelihood = -187.26756                  Pseudo R2     =   0.2607


------------------------------------------------------------------------------
 capsule |     Coef.    Std. Err.      z     P>|z|     [ 95% Conf. Interval]
---------+--------------------------------------------------------------------
     age | -.0118068    .0197265    -0.599   0.549     -.05047     .0268565
    race |  -.651368    .4721849     1.379   0.168    1.576833     .2740974
 dpros_2 |  .7301866    .3589999     2.034   0.042    .0265597    1.433814
 dpros_3 |  1.509485    .3772037     4.002   0.000    .7701797    2.248791
 dpros_4 |  1.387175    .4620253     3.002   0.003    .4816219    2.292728
   dcaps |  .4923895    .4635987     1.062   0.288   -.4162472    1.401026
     psa |  .0298877    .0100993     2.959   0.003    .0100934    .0496819
 gleason |  .9625094    .166819      5.770   0.000    .6355502    1.289469
     vol |  -.011462    .0078144    -1.467   0.142    -.026778    .0038541
   _cons | -6.807546    1.737906    -3.917   0.000   -10.21378   -3.401312
------------------------------------------------------------------------------
```

```
. logit capsule age race dpros_2 dpros_3 dpros_4 dcapsnew psa gleason vol

Iteration 0:   Log Likelihood =-253.29367
Iteration 1:   Log Likelihood =-193.46088
Iteration 2:   Log Likelihood =-187.54897
Iteration 3:   Log Likelihood =-187.26856
Iteration 4:   Log Likelihood =-187.26756

Logit Estimates                              Number of obs =      376
                                             chi2(9)       =   132.05
                                             Prob > chi2   =   0.0000
Log Likelihood = -187.26756                  Pseudo R2     =   0.2607


------------------------------------------------------------------------------
 capsule |     Coef.    Std. Err.      z     P>|z|     [95% Conf. Interval]
---------+--------------------------------------------------------------------
     age | -.0118068    .0197265    -0.599   0.549     -.05047     .0268565
    race |  -.651368    .4721849    -1.379   0.168   -1.576833     .2740974
 dpros_2 |  .7301866    .3589999     2.034   0.042    .0265597    1.433814
 dpros_3 |  1.509485    .3772037     4.002   0.000    .7701797    2.248791
 dpros_4 |  1.387175    .4620253     3.002   0.003    .4816219    2.292728
dcapsnew |  .4923895    .4635987     1.062   0.288   -.4162472    1.401026
     psa |  .0298877    .0100993     2.959   0.003    .0100934    .0496819
 gleason |  .9625094    .166819      5.770   0.000    .6355502    1.289469
     vol |  -.011462    .0078144    -1.467   0.142    -.026778    .0038541
   _cons | -6.315156    1.713852    -3.685   0.000   -9.674244   -2.956068
------------------------------------------------------------------------------
```

The variable DCAPS does not need to be recoded. The tables above show that the estimated coefficients obtained from fitting a model containing DCAPS and from one using a 0-1 coded design variable DCAPSNEW are the same, with the exception of the coefficient for the constant term. Although, recoding is not essential, for ease of interpretation the design variable will be used in future analyses using this variable. Also RACE will be recoded as a 0-1 design variable, RACENEW (0=White, 1=Black) in future analyses.

(c)   *Write down the equation for the logistic regression model of CAPSULE on AGE, RACE DPROS, DCAPS, PSA, GLEASON and VOL. Write down the equation for the logit transformation of this logistic regression model. How many parameters does this model contain?*

logistic regression model:

$$\pi(\mathbf{x}) = \frac{e^{\beta_0 + \beta_1*(AGE) + \beta_2*RACENEW + \beta_3*(DPROS\_2) + \beta_4*(DPROS\_3) + \beta_5*(DPROS\_4) + \beta_6*(DCAPSNEW) + \beta_7*(PSA) + \beta_8*(GLEASON) + \beta_9*(VOL)}}{1 + e^{\beta_0 + \beta_1*(AGE) + \beta_2*RACENEW + \beta_3*(DPROS\_2) + \beta_4*(DPROS\_3) + \beta_5*(DPROS\_4) + \beta_6*(DCAPSNEW) + \beta_7*(PSA) + \beta_8*(GLEASON) + \beta_9*(VOL)}}$$

where $\mathbf{x} = $ set of covariates

logit transformation:

$$g(\mathbf{x}) = \beta_0 + \beta_1 * (AGE) + \beta_2 * (RACENEW) + \beta_3 * (DPROS\_2) + \beta_4 * (DPROS\_3) + \beta_5 * (DPROS\_4) + \beta_6 * (DCAPSNEW) + \beta_7 * (PSA) + \beta_8 * (GLEASON) + \beta_9 * (VOL)$$

This model contains 10 parameters.

(d)   *Write down an expression for the likelihood and log likelihood for the logistic regression model in Exercise 2(c). How many likelihood equations are there? Write down an expression for a typical likelihood equation for this problem.*

likelihood function:

$$l(\beta) = \prod_{i=1}^{n} \zeta(\mathbf{x}_i) \qquad \text{where, } \zeta(\mathbf{x}_i) = \pi(\mathbf{x}_i)^{y_i}\left[1 - \pi(\mathbf{x}_i)\right]^{1-y_i}$$

$\mathbf{x} = $ vector of covariates   and $y_i = \begin{cases} 0 & \text{no penetration} \\ 1 & \text{penetration} \end{cases}$

log likelihood function:

$$L(\beta) = \ln\left[l(\beta)\right] = \sum_{i=1}^{n}\left\{y_i \ln\left[\pi(\mathbf{x}_i)\right] + (1 - y_i)\ln\left[1 - \pi(\mathbf{x}_i)\right]\right\}$$

There will be $p + 1$ or 11 likelihood equations for this problem.

Likelihood equations that result may be expressed as follows:

$$\sum_{i=1}^{n}\left[y_i - \pi(\mathbf{x}_i)\right] = 0$$

and

$$\sum_{i=1}^{n}x_{ij}\left[y_i - \pi(\mathbf{x}_i)\right] = 0, \quad j = 1,\dots,p$$

(e)   *Using a logistic regression package, obtain the maximum likelihood estimates of the parameters of the logistic regression model in Exercise 2(c). Using these estimates, write down the equation for the fitted values, that is the estimated logistic probabilities.*

```
. logit capsule age racenew dpros_2 dpros_3 dpros_4 dcapsnew psa gleason vol

Iteration 0:   Log Likelihood =-253.29367
Iteration 1:   Log Likelihood =-193.46088
Iteration 2:   Log Likelihood =-187.54897
Iteration 3:   Log Likelihood =-187.26856
Iteration 4:   Log Likelihood =-187.26756

Logit Estimates                                Number of obs =      376
                                               chi2(9)       = 132.05
                                               Prob > chi2   = 0.0000
Log Likelihood = -187.26756                    Pseudo R2     = 0.2607

--------------------------------------------------------------------------
 capsule |     Coef.   Std. Err.      z     P>|z|    [95% Conf. Interval]
---------+----------------------------------------------------------------
     age | -.0118068   .0197265   -0.599   0.549     -.05047    .0268565
 racenew |  -.651368   .4721849   -1.379   0.168   -1.576833    .2740974
 dpros_2 |  .7301866   .3589999    2.034   0.042    .0265597    1.433814
 dpros_3 |  1.509485   .3772037    4.002   0.000    .7701797    2.248791
 dpros_4 |  1.387175   .4620253    3.002   0.003    .4816219    2.292728
dcapsnew |  .4923895   .4635987    1.062   0.288   -.4162472    1.401026
     psa |  .0298877   .0100993    2.959   0.003    .0100934    .0496819
 gleason |  .9625094    .166819    5.770   0.000    .6355502    1.289469
     vol |  -.011462   .0078144   -1.467   0.142    -.026778    .0038541
   _cons | -6.966524   1.619597   -4.301   0.000   -10.14088   -3.792172
--------------------------------------------------------------------------
```

logit transformation

$$g(x) = -6.967 + (-.012)*(AGE) + (-.651)*(RACENEW) + (.730)*(DPROS\_2) + (1.509)*(DPROS\_3) +$$
$$(1.387)*(DPROS\_4) + (.492)*(DCAPSNEW) + (.030)*(PSA) + (.963)*(GLEASON) + (-.011)*(VOL)$$

logistic regression model:

$$\pi(x) = \frac{e^{\substack{-6.967+(-.012)*(AGE)+(-.651)*(RACENEW)+(.730)*(DPROS\_2)+(1.509)*(DPROS\_3)+(1.387)*(DPROS\_4)+(.492)*(DCAPSNEW) \\ +(.030)*(PSA)+(.963)*(GLEASON)+(-.011)*(VOL)}}}{1 + e^{\substack{-6.967+(-.012)*(AGE)+(-.651)*(RACENEW)+(.730)*(DPROS\_2)+(1.509)*(DPROS\_3)+(1.387)*(DPROS\_4)+(.492)*(DCAPSNEW) \\ +(.030)*(PSA)+(.963)*(GLEASON)+(-.011)*(VOL)}}}$$

(f)   *Using the results of the output from the logistic regression package used in Exercise 2(e), assess the significance of the slope coefficients for the variables in the model using the likelihood ratio test. What assumptions are needed for the p-values computed for this test to be valid? What is the value of the deviance for the fitted model?*

Deviance:

$$D = -2 \ln\left[\frac{(\text{likelihood of the current model})}{(\text{likelihood of the saturated model})} \right]$$

$$D = -2(-187.26756)$$

$$D = 374.53512$$

Likelihood Ratio Test:

$$H_0 : \beta_1 = \beta_2 = \beta_3 = \beta_4 = \beta_5 = \beta_6 = \beta_7 = \beta_8 = 0$$

H_A: At least one coefficient is not equal to 0

$$G = D(\text{model without variable}) - D(\text{model with variable})$$

$$G = 512.28884 - 374.53512$$

$$G = 137.75372 \qquad\qquad G \sim \chi^2(9) \qquad p < 0.0001$$

∴ reject H_0, it is not consistent with the data that all $\beta=0$; we conclude that together, AGE, RACE, DPROS, DCAPS, PSA, GLEASON and VOL are significant predictors of CAPSULE.

• <u>Assumption</u>: the statistic G will follow a χ^2 distribution with 9 degrees of freedom under the null hypothesis.

The value of the deviance for the fitted model is: $D = 374.54$.

(g) *Use the Wald statistics to obtain an approximation to the significance of the individual slope coefficients for the variables in the model. Fit a reduced model that eliminates those variables with nonsignificant Wald statistics. Assess the joint (conditional) significance of the variables excluded from the model. Present the results of fitting the reduced model in a table.*

Using a critical value of p<0.05, based on the computer output from Exercise 2(e), one can conclude that the variables DPROS_2, DPROS_3, DPROS_4, PSA and GLEASON are significant while AGE, DCAPSNEW, and VOL are not significant. A reduced model was fit containing only those variables thought to be significant:

```
logit capsule dpros_2 dpros_3 dpros_4 psa gleason

Iteration 0:   Log Likelihood =-256.14442
Iteration 1:   Log Likelihood =-196.99369
Iteration 2:   Log Likelihood =-191.28391
Iteration 3:   Log Likelihood =-191.06178
Iteration 4:   Log Likelihood =-191.06126

Logit Estimates                              Number of obs =      380
                                             chi2(5)       = 130.17
                                             Prob > chi2   = 0.0000
Log Likelihood = -191.06126                  Pseudo R2     = 0.2541

------------------------------------------------------------------------
capsule |    Coef.    Std. Err.      z     P>|z|    [95% Conf. Interval]
--------+---------------------------------------------------------------
dpros_2 |  .7656113   .3564098     2.148   0.032    .0670609    1.464162
dpros_3 |  1.562414   .3716103     4.204   0.000    .8340715    2.290757
dpros_4 |  1.439938   .4490547     3.207   0.001    .5598069    2.320069
    psa |  .0280247   .0093822     2.987   0.003    .0096359    .0464136
gleason |  .9995365   .1612634     6.198   0.000    .6834661    1.315607
  _cons | -8.183523   1.057289    -7.740   0.000   -10.25577   -6.111274
--------+---------------------------------------------------------------
```

The likelihood ratio test comparing the above model with the full model will have a distribution that is chi-square with 4 degrees of freedom under the hypothesis that the coefficients for the variables that were excluded are equal to zero.

Likelihood Ratio Test:

H_0: Coeffients for eliminated variables all equal 0

H_A: At least one coefficient is not equal to 0

$G = D(\text{model without variables}) - D(\text{model with variables})$

$$G = 382.12252 - 374.53512$$

$$G = 7.5874 \qquad G \sim \chi^2(4) \qquad p = 0.10792$$

$\therefore$ do not reject H_0, it is consistent with the data that the coefficients for the eliminated variables are all equal to zero; we conclude that the reduced model containing only DPROS, PSA and GLEASON is as good as the full model. Statistically speaking, there is no advantage to including AGE, RACE, DCAPSNEW or VOL in the model. The decision to include any or all of these variables should be made after considering clinical reasons for their inclusion in the model.

Estimated Coefficients for a Multiple Logistic Regression Model for CAPSULE using the Variables DPROS, PSA and GLEASON from the Prostate Study (n=380)

Variable	Coeff.	Std.Err.	z	P>|z|
DPROS_2	0.766	0.3564	2.15	0.032
DPROS_3	1.562	0.3716	4.20	<0.001
DPROS_4	1.440	0.4491	3.21	0.001
PSA	0.028	0.0094	2.99	0.003
GLEASON	1.000	0.1613	6.20	<0.001
Constant	-8.184	1.0573	-7.74	<0.001
Log likelihood = -191.061				

(h)   *Using the results from Exercise 2(g), compute 95 percent confidence intervals for all coefficients in the model. Write a sentence interpreting the confidence intervals for the non-constant covariates.*

Ninety-five percent confidence intervals for the coefficients for DPROS_2, DPROS_3, DPROS_4, PSA and GLEASON can be computed using equations (1.15) and (1.16) from the text. The intervals shown below are calculated from the STATA output for the logistic regression model shown in Exercise 2(g). The confidence intervals for the coefficients can also be seen in the 6th and 7th columns presented in this STATA output.

Endpoints of a $100(1-\alpha)\%$ confidence interval for DPROS_2:

$$\hat{\beta}_1 \pm z_{1-\alpha/2} S\hat{E}\left(\hat{\beta}_1\right)$$

$$0.7656 \pm 1.96(0.3564)$$

$$(0.0671, 1.4642)$$

Endpoints of a $100(1-\alpha)\%$ confidence interval for DPROS_3:

$$\hat{\beta}_2 \pm z_{1-\alpha/2} S\hat{E}(\hat{\beta}_2)$$

$$1.5624 \pm 1.96(0.3716)$$

$$(0.8341, 2.2908)$$

Endpoints of a $100(1-\alpha)\%$ confidence interval for DPROS_4:

$$\hat{\beta}_3 \pm z_{1-\alpha/2} S\hat{E}(\hat{\beta}_3)$$

$$1.4400 \pm 1.96(0.4491)$$

$$(0.5598, 2.3201)$$

Endpoints of a $100(1-\alpha)\%$ confidence interval for PSA:

$$\hat{\beta}_4 \pm z_{1-\alpha/2} S\hat{E}(\hat{\beta}_4)$$

$$0.0280 \pm 1.96(0.0094)$$

$$(0.0096, 0.0464)$$

Endpoints of a $100(1-\alpha)\%$ confidence interval for GLEASON:

$$\hat{\beta}_5 \pm z_{1-\alpha/2} \hat{SE}\left(\hat{\beta}_5\right)$$

$$0.9995 \pm 1.96(0.1613)$$

$$(0.6835, 1.3156)$$

Endpoints of a $100(1-\alpha)\%$ confidence interval for the constant:

$$\hat{\beta}_0 \pm z_{1-\alpha/2} \hat{SE}\left(\hat{\beta}_0\right)$$

$$-8.1834 \pm 1.96(1.0573)$$

$$(-10.2558, -6.1113)$$

The 95% confidence interval for DPROS_2 suggests that the change in the log odds of capsule penetration (CAPSULE=1) when there is a unilobar nodule on the left (compared with no nodule) is 0.7656 when the values of PSA and GLEASON are constant and that the change could be as little as 0.0671 or as much as 1.4642 with 95% confidence.

The 95% confidence interval for DPROS_3 suggests that the change in the log odds of capsule penetration (CAPSULE=1) when there is a unilobar nodule on the right (compared with no nodule) is 1.5624 when the values of PSA and GLEASON are constant and that the change could be as little as 0.8341 or as much as 2.2908 with 95% confidence.

The 95% confidence interval for DPROS_4 suggests that the change in the log odds of capsule penetration (CAPSULE=1) when there is a bilobar nodule (compared with no nodule) is 1.4399 when the values of PSA and GLEASON are constant and that the change could be as little as 0.5598 or as much as 2.3201 with 95% confidence.

The 95% confidence interval for PSA suggests that the change in the log odds of capsule penetration (CAPSULE=1) per one unit change in PSA is 0.0280 when the values of DPROS and GLEASON are constant and that the change could be as little as 0.0096 or as much as 0.0464 with 95% confidence.

The 95% confidence interval for GLEASON suggests that the change in the log odds of capsule penetration (CAPSULE=1) per one unit change in GLEASON is 0.9995 when the values of DPROS and PSA are constant and that the change could be as little as 0.6835 or as much as 1.3156 with 95% confidence.

(i)   *Obtain the estimated covariance matrix for the final model fit in Exercise 2(h). Choose a set of values for the covariates in that model and estimate the logit and logistic probability for the population of subjects with these characteristics. Compute 95 percent confidence intervals for the logit and logistic probability. Write a sentence or two interpreting the estimated probability and its confidence interval.*

```
. corr, covariance _coef

          |  dpros_2   dpros_3   dpros_4       psa   gleason      _cons
----------+---------------------------------------------------------------
  dpros_2 |  .127028
  dpros_3 |  .084943   .138094
  dpros_4 |  .082651   .081982    .20165
      psa |  .000323   .000483  -.000203   .000088
  gleason | -.002756  -.000054  -.001146  -.000414   .026006
    _cons | -.070627  -.090788  -.072543   .001256  -.162309   1.11786
```

For a patient who has a bilobar node, a PSA value of 9.5 and a GLEASON score of 4:

Estimated logit:

$$g(x) = \beta_0 + \beta_1 * (DPROS\_2) + \beta_2 * (DPROS\_3) + \beta_3 * (DPROS\_4) + \beta_4 * (PSA) + \beta_5 * (GLEASON)$$

$$= (-8.184) + (1.440) * (1) + (0.028) * (9.5) + (1.000) * (4)$$

$$= -2.478$$

$$Var[\hat{g}(x)] = 1.11786 + (1)^2 * (0.20165) + (9.5)^2 * (0.000088) + (4)^2 * (0.026006)$$
$$+ 2 * 1 * 9.5 * (-0.000203) + 2 * 1 * 4 * (-0.001146) + 2 * 9.5 * 4 * (-0.000414)$$
$$+ 2 * 1 * (-0.072543) + 2 * 9.5 * (0.001256) + 2 * 4 * (-0.162309)$$

$$= 0.279365$$

$$SE[\hat{g}(x)] = 0.5285$$

The estimated standard error of the logit for a subject with a bilobar nodes, a PSA of 9.5 and a Gleason score of 4, can be used to calculate the endpoints of the confidence interval for the logit for a subject with a bilobar nodes, a PSA of 9.5 and a Gleason score of 4:

$$\hat{g}(x) \pm z_{1-\alpha/2} SE\left[\hat{g}(x)\right]$$

$$-2.478 \pm 1.96 * 0.5285$$

$$(-3.514, -1.442)$$

Estimated logistic probability:

$$\hat{\pi}(x) = \frac{e^{\hat{g}(x)}}{1 + e^{\hat{g}(x)}}$$

$$= \frac{e^{(-8.184)+(1.440)*(1)+(0.028)*(9.5)+(1.000)*(4)}}{1 + e^{(-8.184)+(1.440)*(1)+(0.028)*(9.5)+(1.000)*(4)}}$$

$$= 0.0774$$

The endpoints of the confidence interval for the estimated logistic probability can be obtained by following a similar process using the endpoints of the estimated logit.

The lower limit is:

$$\frac{e^{-3.514}}{1 + e^{-3.514}} = 0.0289$$

And the upper limit is:

$$\frac{e^{-1.442}}{1 + e^{-1.442}} = 0.1912$$

The estimated logistic probability of tumor invasion of the prostatic capsule for a men with bilobar nodes, PSA of 9.5 and a Gleason score of 4, 0.0774, is an estimate of the proportion of men with these characteristics in the population sampled with tumor invasion of the prostatic capsule. The confidence interval suggests that this mean could be as low as 0.0289 or as high as 0.1912 with 95% confidence.

Chapter Three – Solutions

1. Consider the ICU data described in Section 1.6.1 and use as the outcome variable vital status (STA) and CPR prior to ICU admission (CPR) as a covariate.

   (a)   Demonstrate that the value of the log-odds ratio obtained from the cross-classification of STA by CPR is identical to the estimated slope coefficient from the logistic regression of STA on CPR. Verify that the estimated standard error of the estimated slope coefficient for CPR obtained from the logistic regression package is identical to the square root of the sum of the inverse of the cell frequencies from the cross-classification of STA by CPR. Use either set of computations to obtain the 95% CI for the odds ratio. What aspect concerning the coding of the variable CPR makes the calculations for the two methods equivalent?

```
. tab  cpr sta

          |        sta
     cpr  |        0           1 |     Total
----------+----------------------+----------
       0  |      154          33 |       187
       1  |        6           7 |        13
----------+----------------------+----------
   Total  |      160          40 |       200
```

$$OR = \frac{a*d}{b*c} = \frac{(154)*(7)}{(33)*(6)} = 5.44$$

$$\ln(OR) = 1.6946$$

```
. logit sta cpr

Iteration 0:   Log Likelihood =-100.08048
Iteration 1:   Log Likelihood =-96.634239
Iteration 2:   Log Likelihood =-96.117589
Iteration 3:   Log Likelihood =-96.114275

Logit Estimates                               Number of obs =      200
                                              chi2(1)       =     7.93
                                              Prob > chi2   =   0.0049
Log Likelihood = -96.114275                   Pseudo R2     =   0.0396

--------------------------------------------------------------------------
     sta |      Coef.   Std. Err.       z     P>|z|    [95% Conf. Interval]
---------+----------------------------------------------------------------
     cpr |   1.694596    .5884886     2.880   0.004     .5411793    2.848012
   _cons |  -1.540445    .1918242    -8.031   0.000    -1.916414   -1.164476
--------------------------------------------------------------------------
```

The log-odds ratio from the 2x2 contingency table is identical to the estimated slope coefficient from the logistic regression of STA on CPR.

Verify that the estimated standard error of the estimated slope coefficient for CPR obtained from the logistic regression package is identical to the square root of the sum of the inverse of the cell frequencies from the cross-classification of STA by CPR.

From the output above, it is clear that the estimated standard error of the estimated slope coefficient from the logistic regression is:

$$S\hat{E}(\hat{\beta_1}) = 0.588$$

This is equivalent to the square root of the sum of the inverse of the cell frequencies from the 2x2 contingency table:

$$S\hat{E}\left(\ln(\hat{OR})\right) = sqrt\left(\frac{1}{154} + \frac{1}{33} + \frac{1}{6} + \frac{1}{7}\right) = 0.588$$

Use either set of computations to obtain 95% CI for the odds ratio.

$$95\%CI = \exp[\hat{\beta_1} \pm z_{1-\alpha/2} * S\hat{E}(\hat{\beta_1})]$$
$$= \exp[1.6946 \pm 1.96 * 0.588]$$

$$1.718 \leq OR \leq 17.237$$

This is equivalent to the 95% CI for the estimated odds ratio that can be obtained from the logistic procedure in STATA:

```
. logistic sta cpr

Logit Estimates                              Number of obs =      200
                                             chi2 (1)      =     7.93
                                             Prob > chi2   = 0.0049
Log Likelihood = -96.114275                  Pseudo R2     = 0.0396


---------------------------------------------------------------------
     sta | Odds Ratio   Std. Err.      z     P>|z|    [95% Conf. Interval]
---------+-----------------------------------------------------------
     cpr |   5.444444    3.203994    2.880   0.004    1.718032   17.25345
---------------------------------------------------------------------
```

What aspect concerning the coding of the variable CPR makes the calculations for the two methods equivalent?

The calculations from the two methods are equivalent because the variable CPR is dichotomous and has been coded 0,1.

(b)   For purposes of illustration, use a data transformation statement to recode, for this problem only, the variable CPR as follows: 4 = no and 2 = yes. Perform the logistic regression of STA on CPR (recoded). Demonstrate how the calculation of the logit difference of CPR = yes versus CPR = no is equivalent to the value of the

log-odds ratio obtained in Exercise 1(a). Use the results from the logistic regression to obtain the 95% CI for the odds ratio and verify that they are the same limits as obtained in Exercise 1(a).

```
. gen cprnew=2
. replace cprnew=4 if cpr==0
(187 real changes made)
```

Perform the logistic regression of STA on CPR (recoded).

```
. logit sta cprnew

Iteration 0:   Log Likelihood =-100.08048
Iteration 1:   Log Likelihood =-96.634239
Iteration 2:   Log Likelihood =-96.117589
Iteration 3:   Log Likelihood =-96.114275

Logit Estimates                                 Number of obs =      200
                                                chi2(1)       =     7.93
                                                Prob > chi2   =   0.0049
Log Likelihood = -96.114275                     Pseudo R2     =   0.0396


---------------------------------------------------------------------------
     sta |    Coef.   Std. Err.      z     P>|z|    [95% Conf. Interval]
---------+-----------------------------------------------------------------
  cprnew |  -.8472979  .2942443   -2.880   0.004   -1.424006    -.2705896
   _cons |   1.048746  1.129108    1.637   0.102    -.3642654    4.061758
---------------------------------------------------------------------------
```

Demonstrate how the calculation of the logit difference of CPR=yes vs. CPR=no is equivalent to the value of the log-odds ratio obtained in Exercise 1(a).

$$\ln\left[\hat{OR}(a,b)\right] = \hat{g}(x=a) - \hat{g}(x=b)$$

$$= \left(\hat{\beta}_0 + \hat{\beta}_1 * a\right) - \left(\hat{\beta}_0 + \hat{\beta}_1 * b\right)$$

$$= \hat{\beta}_1 * (a-b)$$

$$= -0.8472979 * (2-4)$$

$$= 1.6946$$

This is equivalent to the value of the log-odds ratio obtained in problem 1(a).

Use the results from the logistic regression to obtain the 95% CI for the odds ratio and verify that they are the same limits as obtained in Exercise 1(a).

$$95\%CI = \exp[2\hat{\beta}_1 \pm z_{1-\alpha/2} * 2S\hat{E}\left(\hat{\beta}_1\right)]$$

Under this coding scheme, the independent variable takes on a lower value (2) when the covariate is present than when the covariate is absent (4). In order to preserve the comparison that was made in Exercise 1(a), the sign of the coefficient is changed from negative to positive. This will lead to the appropriate confidence interval for the odds ratio.

$$95\%CI = \exp[2*(0.8472979) \pm 1.96*2*(0.2942443)]$$

$$= \exp[1.6946 \pm 1.96*0.588]$$

$$1.718 \le OR \le 17.237$$

This is the same confidence interval for the odds ratio that was obtained in Exercise 1(a).

(c)   *Consider the ICU data and use as the outcome variable vital status (STA) and race (RACE) as a covariate. Prepare a table showing the coding of the two design variables for RACE using the value RACE = 1, white, as the reference group. Show that the estimated log-odds ratios obtained from the cross-classification of STA by RACE, using RACE = 1 as the reference group, are identical to estimated slope coefficients for the two design variables from the logistic regression of STA on RACE. Verify that the estimated standard errors of the estimated slope coefficients for the two design variables for RACE are identical to the square root of the sum of the inverse of the cell frequencies from the cross-classification of STA by RACE used to calculate the odds ratio. Use either set of computations to compute the 95% CI for the odds ratios.*

RACE	Label	RACE_2	RACE_3
1	White	0	0
2	Black	1	0
3	Other	0	1

Show that the estimated log-odds ratios obtained from the cross-classification of STA by RACE, using RACE=1 as the reference group, are identical to estimated slope coefficients for the two design variables from the logistic regression of STA on RACE.

```
. tab race sta

          | sta
     race |         0         1 |     Total
----------+----------------------+----------
        1 |       138        37 |       175
        2 |        14         1 |        15
        3 |         8         2 |        10
----------+----------------------+----------
    Total |       160        40 |       200
```

$$OR(\text{Race} = 2 \text{ vs. Race} = 1) = \frac{a * d}{b * c} = \frac{(138) * (1)}{(37) * (14)} = 0.266$$

$$\ln\big(OR(2 \text{ vs. } 1)\big) = -1.3227$$

$$OR(\text{Race} = 3 \text{ vs. Race} = 1) = \frac{a * d}{b * c} = \frac{(138) * (2)}{(37) * (8)} = 0.9324$$

$$\ln\big(OR(3 \text{ vs. } 1)\big) = -0.069958$$

```
. logit sta r_2 r_3

Iteration 0:   Log Likelihood =-100.08048
Iteration 1:   Log Likelihood =-99.043397
Iteration 2:   Log Likelihood =-98.952555
Iteration 3:   Log Likelihood = -98.95055
Iteration 4:   Log Likelihood =-98.950549

Logit Estimates                         Number of obs =      200
                                        chi2(2)       =     2.26
                                        Prob > chi2   = 0.3231
Log Likelihood = -98.950549             Pseudo R2     = 0.0113

------------------------------------------------------------------------
   sta |     Coef.   Std. Err.       z     P>|z|    [95% Conf. Interval]
-------+----------------------------------------------------------------
   r_2 | -1.322722   1.051523    -1.258    0.208    -3.383669    .7382257
   r_3 | -.0699586    .8119565    -0.086   0.931    -1.661364    1.521447
 _cons | -1.316336    .1851308    -7.110   0.000    -1.679185   -.9534861
------------------------------------------------------------------------
```

The estimated log-odds ratios obtained from the cross-classification of STA by RACE, using RACE=1 as the reference group, are identical to estimated slope coefficients for the two design variables from the logistic regression of STA on RACE.

Verify that the estimated standard errors of the estimated slope coefficients for the two design variables for RACE are identical to the square root of the sum of the inverse of the cell frequencies from the cross-classification of STA by RACE used to calculate the odds ratio.

For RACE 2 vs. RACE 1

$$\hat{SE}(\hat{\beta}_1) = 1.051523 \quad \text{from the estimated logit}$$

$$\hat{SE}\big(\ln(\hat{OR})\big) = sqrt\left(\frac{1}{138} + \frac{1}{1} + \frac{1}{37} + \frac{1}{14}\right) = 1.051523$$

For RACE 3 vs. RACE 1

$$\hat{SE}(\hat{\beta}_1) = 0.8119565 \quad \text{from the estimated logit}$$

$$\hat{SE}\big(\ln(\hat{OR})\big) = sqrt\left(\frac{1}{138} + \frac{1}{2} + \frac{1}{37} + \frac{1}{8}\right) = 0.8119565$$

Use either set of computations to compute the 95% CI for the odds ratios.

For RACE 2 vs. RACE 1

$$95\%CI = \exp[\hat{\beta}_1 \pm z_{1-\alpha/2} * \hat{SE}(\hat{\beta}_1)]$$

$$= \exp[-1.3227 \pm 1.96 * (1.051523)]$$

$$0.03392 \leq OR \leq 2.0923$$

For RACE 3 vs. RACE 1

$$95\%CI = \exp[\hat{\beta}_1 \pm z_{1-\alpha/2} * \hat{SE}(\hat{\beta}_1)]$$

$$= \exp[-0.0699586 \pm 1.96 * (0.811956)]$$

$$0.18987 \leq OR \leq 4.578979$$

These are equivalent to the 95% confidence intervals for the estimated odds ratios that can be obtained from the logistic procedure in STATA:

```
. logistic sta r_2 r_3

Logit Estimates                                 Number of obs =      200
                                                chi2(2)       =     2.26
                                                Prob > chi2   = 0.3231
Log Likelihood = -98.950549                     Pseudo R2     = 0.0113

--------------------------------------------------------------------------
    sta | Odds Ratio   Std. Err.       z     P>|z|    [95% Conf. Interval]
--------+-----------------------------------------------------------------
    r_2 |   .2664093    .2801355    -1.258    0.208     .0339228    2.09222
    r_3 |   .9324324    .7570946    -0.086    0.931     .1898798   4.578846
--------------------------------------------------------------------------
```

(d)   Create design variables for RACE using the method typically employed in ANOVA. Perform the logistic regression of STA on RACE. Show by calculation that the estimated logit differences of RACE = 2 versus RACE = 1 and RACE = 3 versus RACE = 1 are equivalent to the values of the log-odds ratio obtained in problem 1(c). Use the results of the logistic regression to obtain the 95% CI for the odds ratios and verify that they are the same limits as obtained in Exercise 1(c). Note that the estimated covariance matrix for the estimated coefficients is needed to obtain the estimated variances of the logit differences.

```
. gen rdvm_1=-1

. replace rdvm_1=1 if race==2
(15 real changes made)

. replace rdvm_1=0 if race==3
(10 real changes made)

. gen rdvm_2=-1

. replace rdvm_2=0 if race==2
(15 real changes made)

. replace rdvm_2=1 if race==3
(10 real changes made)
```

RACE	Label	RDVM_1	RDVM_2
1	White	-1	-1
2	Black	1	0
3	Other	0	1

Perform the logistic regression of STA on RACE.

```
. logit

Logit Estimates                              Number of obs =      200
                                             chi2(2)       =     2.26
                                             Prob > chi2   =   0.3231
Log Likelihood = -98.950549                  Pseudo R2     =   0.0113

------------------------------------------------------------------------
    sta |      Coef.   Std. Err.       z      P>|z|    [95% Conf. Interval]
--------+---------------------------------------------------------------
 rdvm_1 |  -.8584948   .7412439    -1.158    0.247    -2.311306    .5943165
 rdvm_2 |   .3942681   .6329561     0.623    0.533     -.846303    1.634839
  _cons |  -1.780562   .4385203    -4.060    0.000    -2.640047   -.9210784
------------------------------------------------------------------------
```

Show by calculation that the estimated logit differences of RACE=2 vs. RACE=1 and RACE=3 vs. RACE=1 are equivalent to the values of the log-odds ratio obtained in problem 2.1.

For RACE 2 vs. RACE 1

$$\ln\left[\hat{OR}(\text{black}, \text{white})\right] = \hat{g}(\text{black}) - \hat{g}(\text{white})$$

$$= \hat{\beta}_0 + \hat{\beta}_{11} * (D_1 = 1) + \hat{\beta}_{12} * \left(D_2 = 0\right)$$
$$- \left[\hat{\beta}_0 + \hat{\beta}_{11} * (D_1 = -1) + \hat{\beta}_{12} * \left(D_2 = -1\right)\right]$$
$$= 2\hat{\beta}_{11} + \hat{\beta}_{12}$$
$$= 2 * (-0.8584948) + 0.3942681$$

$$\ln\left[\hat{OR}(\text{black}, \text{white})\right] = -1.322722$$

For RACE 3 vs. RACE 1

$$\ln\left[\hat{OR}(\text{other}, \text{white})\right] = \hat{g}(\text{other}) - \hat{g}(\text{white})$$

$$= \hat{\beta}_0 + \hat{\beta}_{11} * (D_1 = 0) + \hat{\beta}_{12} * (D_2 = 1)$$

$$- \left[\hat{\beta}_0 + \hat{\beta}_{11} * (D_1 = -1) + \hat{\beta}_{12} * (D_2 = -1)\right]$$

$$= \hat{\beta}_{11} + 2\hat{\beta}_{12}$$

$$= (-0.8584948) + 2 * (0.3942681)$$

$$= -0.0699586$$

Use the results of the logistic regression to obtain 95% confidence intervals for the ORs and verify that they are the same limits as obtained in problem 1(c).

```
. correlate, _coef covariance

        |    rdvm_1    rdvm_2      _cons
--------+-------------------------------
 rdvm_1 |   .549443
 rdvm_2 |  -.373176   .400633
  _cons |   .164842   .016033      .1923
```

For RACE 2 vs. RACE 1

$$\hat{Var}\left\{\ln\left[\hat{OR}(\text{black, white})\right]\right\} = 4\,\hat{var}(\hat{\beta}_{11}) + \hat{var}(\hat{\beta}_{12}) + 4\,\hat{cov}(\hat{\beta}_{11}, \hat{\beta}_{12})$$

$$= 4(0.549443) + (0.400633) + 4(-0.373176)$$

$$= 1.105701$$

$$\hat{SE} = \sqrt{1.105701} = 1.0515$$

$$95\%CI = \exp[2\hat{\beta}_{11} + \hat{\beta}_{12} \pm z_{1-\alpha/2} * \hat{SE}(2\hat{\beta}_{11} + \hat{\beta}_{12})]$$

$$= \exp[-1.3227 \pm 1.96 * (1.051523)]$$

$$0.03392 \le OR \le 2.0923$$

For RACE 3 vs. RACE 1

$$\hat{Var}\left\{\ln\left[\hat{OR}(\text{other, white})\right]\right\} = \hat{var}\left(\hat{\beta}_{11}\right) + 4\,\hat{var}\left(\hat{\beta}_{12}\right) + 4\,\hat{cov}\left(\hat{\beta}_{11}, \hat{\beta}_{12}\right)$$

$$= (0.549443) + 4(0.400633) + 4(-0.373176)$$

$$= 0.659271$$

$$\hat{SE} = \sqrt{0.659271} = 0.811955048$$

$$95\%\text{CI} = \exp[\hat{\beta}_{11} + 2\hat{\beta}_{12} \pm z_{1-\alpha/2} * \hat{SE}\left(\hat{\beta}_{11} + 2\hat{\beta}_{12}\right)]$$

$$= \exp[-0.0699586 \pm 1.96 * (0.811956)]$$

$$0.18987 \le OR \le 4.578979$$

These are the same confidence intervals as the ones obtained in Exercise 1(c).

(e)   *Consider the variable AGE in the ICU data set. Prepare a table showing the coding of three design variables based on the empirical quartiles of AGE using the first quartile as the reference group. Fit the logistic regression of STA on AGE as recoded into these design variables and plot the three estimated slope coefficients versus the midpoint of the respective age quartile. Plot as a fourth point a value of zero at the midpoint of the first quartile of age. Does the plot suggest that the logit is linear in age?*

AGEGP4	Interval	Midpoint	AQ_2	AQ_3	AQ_4
1	16-46	31.5	0	0	0
2	47-63	55.5	1	0	0
3	64-72	68.5	0	1	0
4	73-92	83.5	0	0	1

Fit the logistic regression of STA on AGE as recoded into these design variables.

```
. logit sta  aq_2 aq_3 aq_4

Logit Estimates                              Number of obs =      200
                                             chi2(3)       =     8.29
                                             Prob > chi2   =   0.0403
Log Likelihood = -95.934112                  Pseudo R2     =   0.0414

--------------------------------------------------------------------------
     sta |     Coef.   Std. Err.       z     P>|z|    [95% Conf. Interval]
---------+----------------------------------------------------------------
    aq_2 |   1.213682   .6160054    1.970    0.049     .0063332    2.42103
    aq_3 |   1.056053   .6298723    1.677    0.094    -.1784743    2.29058
    aq_4 |   1.584897   .6111225    2.593    0.010     .3871187   2.782675
   _cons |  -2.442347    .521286   -4.685    0.000    -3.464049  -1.420645
--------------------------------------------------------------------------
```

Plot the three estimated slope coefficients vs. the midpoint of the respective age quartile. Plot as a fourth point the value zero at the midpoint of the first quartile of age.

Scatterplot of estimated slope coefficient (b) vs. midpoint of age quartile

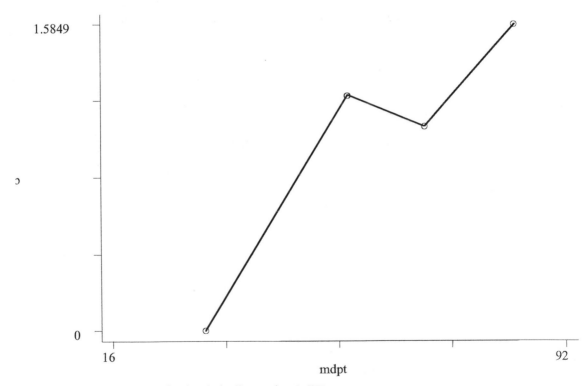

The plot suggests that the logit is linear in AGE.

(f)   *Consider the logistic regression of STA on CRN and AGE. Consider CRN to be the risk factor and show that AGE is a confounder of the association of CRN with STA. Addition of the interaction of AGE by CRN presents an interesting modeling dilemma. Examine the main effects only and interaction models graphically. Using the graphical results and any significance tests you feel are needed, select the best model (main effects or interaction) and justify your choice. Estimate relevant odds ratios. Repeat this analysis of confounding and interaction for a model that includes CPR as the risk factor and AGE as the potential confounding variable.*

```
. logit sta crn

Logit Estimates                              Number of obs =      200
                                             chi2(1)       =     5.42
                                             Prob > chi2   =   0.0199
Log Likelihood = -97.368374                  Pseudo R2     =   0.0271

------------------------------------------------------------------------
    sta |    Coef.    Std. Err.      z     P>|z|    [95% Conf. Interval]
--------+---------------------------------------------------------------
    crn |  1.219757    .5038556    2.421   0.015    .2322178    2.207296
  _cons | -1.53821     .1948369   -7.895   0.000   -1.920084   -1.156337
------------------------------------------------------------------------

. logit sta crn age

Logit Estimates                              Number of obs =      200
                                             chi2(2)       =    11.56
                                             Prob > chi2   =   0.0031
Log Likelihood = -94.302294                  Pseudo R2     =   0.0577

------------------------------------------------------------------------
    sta |    Coef.    Std. Err.      z     P>|z|    [95% Conf. Interval]
--------+---------------------------------------------------------------
    crn |  1.019856    .5149228    1.981   0.048    .0106262    2.029087
    age |  .0249915    .0107232    2.331   0.020    .0039744    .0460085
  _cons | -3.029875    .7000099   -4.328   0.000    4.40107    -1.657881
------------------------------------------------------------------------
```

There is a 19.6% decrease in the value for the coefficient for CRN when AGE is adjusted for in the model. A decrease of this magnitude indicates that AGE confounds the relationship between CRN and STA.

Addition of the interaction of AGE by CRN presents an interesting modeling dilemma.

```
. gen crnage=crn*age
. logit sta crn age crnage

Iteration 0:   Log Likelihood =-100.08048
Iteration 1:   Log Likelihood =-94.160869
Iteration 2:   Log Likelihood =-93.683315
Iteration 3:   Log Likelihood =-93.681076
Iteration 4:   Log Likelihood =-93.681076

Logit Estimates                              Number of obs =      200
                                             chi2(3)       =    12.80
                                             Prob > chi2   =   0.0051
Log Likelihood = -93.681076                  Pseudo R2     =   0.0639

------------------------------------------------------------------------
    sta |    Coef.    Std. Err.      z     P>|z|    [95% Conf. Interval]
--------+---------------------------------------------------------------
    crn |  3.573101   2.322261     1.539   0.124   -.9784469    8.124649
    age |  .029242    .011725      2.494   0.013    .0062613    .0522226
 crnage | -.0380925   .0340579    -1.118   0.263   -.1048447    .0286598
  _cons | -3.297927   .7705345    -4.280   0.000   -4.808147   -1.787707
------------------------------------------------------------------------
```

Examine the main effects only and interaction models graphically.

Scatterplot of the main effects model for STA on CRN AGE
logit vs. age by CRN
logicrn0=logit when CRN=0
logicrn1=logit when CRN=1

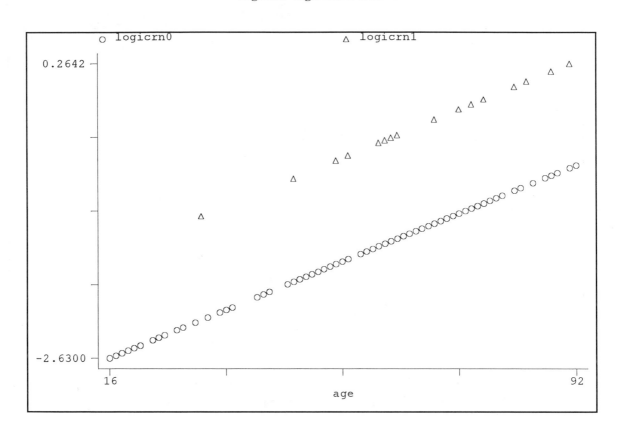

Scatterplot of the interaction model for STA on CRN AGE CRN*AGE
logit vs. AGE by CRN
intcrn0 - logit when CRN=0
intcrn1 - logit when CRN=1

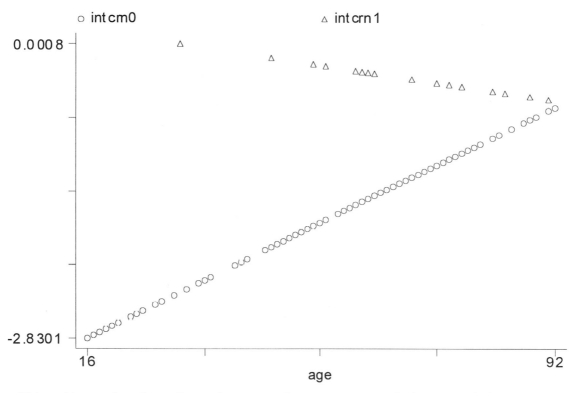

Using the graphical results and any significance tests you feel are needed, select the best model (main effects or interaction) and justify your choice. Estimate the relevant odds ratios.

Model	Constant	crn	age	crn*age	log-likelihood	G	p-value
1	-1.53821	1.219757	-		-97.34		
2	-3.029875	1.019856	0.0249915	-	-94.30	6.08	0.014
3	-3.297927	3.573101	0.029242	-0.038093	-93.68	1.24	0.265

Based on the impression gained from looking at the graph of the logits from the model containing no interaction term, as well as the Wald statistic for the interaction term (crnage) and the results of the likelihood ratio test, it does not appear justified to include the interaction term in the model. There is no effect modification by age.

The relevant odds ratio (adjusted for age) can be obtained by exponentiating the coefficient for CRN from the model that includes AGE as a covariate:

$$\hat{OR} = \exp(1.019856) = 2.77$$

The 95% CI for this odds ratio can be obtained by exponentiating the endpoints of the 95% confidence interval for the coefficient for CRN from the model that includes AGE as a covariate:

$$\exp(0.0106262) \le OR \le \exp(2.029087)$$

$$1.01 \le OR \le 7.61$$

Repeat this analysis of confounding and interaction for a model which includes CPR as the risk factor and AGE as the potential confounding variable.

```
. logit sta cpr

Logit Estimates                                    Number of obs =      200
                                                   chi2(1)       =     7.93
                                                   Prob > chi2   = 0.0049
Log Likelihood = -96.114275                        Pseudo R2     = 0.0396


------------------------------------------------------------------------------
     sta |      Coef.   Std. Err.       z      P>|z|    [95% Conf. Interval]
---------+--------------------------------------------------------------------
     cpr |   1.694596   .5884886      2.880    0.004     .5411793    2.848012
   _cons |  -1.540445   .1918242     -8.031    0.000    -1.916414   -1.164476
------------------------------------------------------------------------------

. logit

Logit Estimates                                    Number of obs =      200
                                                   chi2(2)       =    16.21
                                                   Prob > chi2   = 0.0003
Log Likelihood =  -91.97634                        Pseudo R2     = 0.0810


------------------------------------------------------------------------------
     sta |      Coef.   Std. Err.       z      P>|z|    [95% Conf. Interval]
---------+--------------------------------------------------------------------
     cpr |   1.784092   .6072971      2.938    0.003     .5938116    2.974373
     age |   .0296074   .0111489      2.656    0.008     .0077559    .0514589
   _cons |  -3.351956   .7454995     -4.496    0.000    -4.813108   -1.890803
------------------------------------------------------------------------------
```

There is a 5.0% increase in the value for the coefficient for CPR when AGE is adjusted for in the model. An increase of this magnitude indicates that AGE probably does not confound the relationship between CPR and STA.

Addition of the interaction of AGE by CPR

```
. gen cprage=cpr*age

. logit sta cpr age cprage

Logit Estimates                                  Number of obs =      200
                                                 chi2(3)       =    19.05
                                                 Prob > chi2   =   0.0003
Log Likelihood = -90.554825                      Pseudo R2     =   0.0952

---------------------------------------------------------------------------
     sta |     Coef.    Std. Err.       z      P>|z|     [95% Conf. Interval]
---------+-----------------------------------------------------------------
     cpr | -3.722935     4.21462    -0.883    0.377    -11.98344    4.537568
     age |  .0247665    .0111663     2.218    0.027     .0028809    .0466521
  cprage |  .0941877    .0708038     1.330    0.183    -.0445852    .2329606
   _cons | -3.041958    .7356889    -4.135    0.000    -4.483882   -1.600034
---------------------------------------------------------------------------
```

Examine the main effects only and interaction models graphically.

Scatterplot of the main effects model for STA on CPR AGE
logit vs. age by CPR
logicpr0=logit when CPR=0
logicpr1=logit when CPR=1

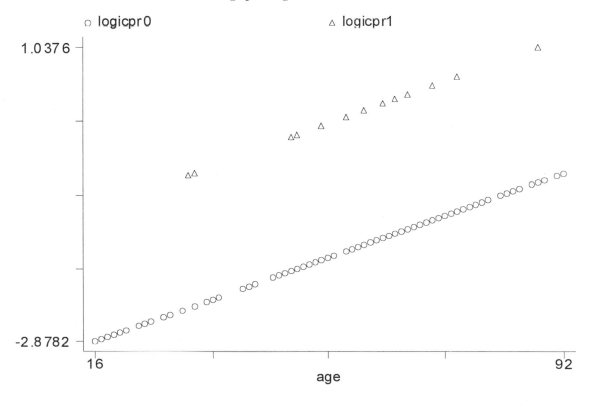

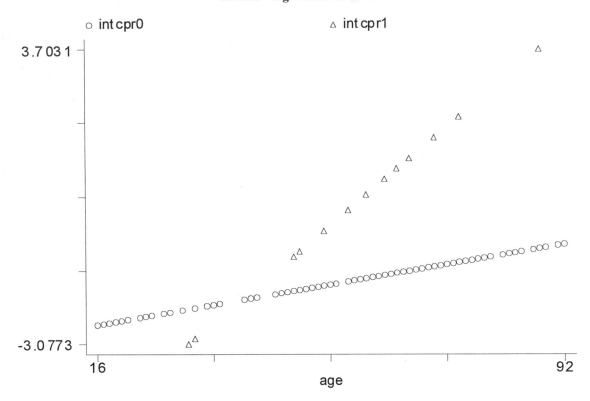

Scatterplot of the interaction model for STA on CPR AGE CPR*AGE
logit vs. AGE by CPR
intcrn0 - logit when CPR=0
intcrn1 - logit when CPR=1

Using the graphical results and any significance tests you feel are needed, select the best model (main effects or interaction) and justify your choice. Estimate the relevant odds ratios.

Model	Constant	cpr	age	cpr*age	log-likelihood	G	p-value
1	-1.540445	1.694596	-	-	-96.11		
2	-3.351956	1.784092	0.0296074	-	-91.98	8.27	0.004
3	-3.041958	-3.722935	0.0247665	0.0941877	-90.55	2.84	0.091

Based on the impression gained from looking at the graph of the logits from the model containing no interaction term, as well as the Wald statistic for the interaction term (cprage) and the results of the likelihood ratio test, it does not appear justified to include the interaction term in the model. There is no effect modification by age.

The relevant odds ratio (crude) can be obtained by exponentiating the coefficient for CPR from the original model

$$\hat{OR} = \exp(1.694596) = 5.44$$

The 95% CI for this odds ratio can be obtained by exponentiating the endpoints of the 95% confidence interval for the coefficient for CPR from the model that includes AGE as a covariate:

$$\exp(0.5938116) \le OR \le \exp(2.974373)$$

$$1.81 \le OR \le 19.58$$

(g)   *Consider an analysis for confounding and interaction for the model with STA as the outcome, CAN as the risk factor, and TYP as the potential confounding variable. Perform this analysis using logistic regression modeling and Mantel-Haenszel analysis. Compare the results of the two approaches.*

CONFOUNDING

Using logistic regression modeling:

```
. logit sta can

Logit Estimates                                   Number of obs =      200
                                                  chi2(1)       =     0.00
                                                  Prob > chi2   =   1.0000
Log Likelihood = -100.08048                       Pseudo R2     =   0.0000


---------------------------------------------------------------------------
     sta |     Coef.    Std. Err.      z      P>|z|     [95% Conf. Interval]
---------+-----------------------------------------------------------------
     can |   1.16e-15    .5892557    0.000    1.000     -1.15492    1.15492
   _cons |  -1.386294    .186339    -7.440    0.000    -1.751512   -1.021077
---------------------------------------------------------------------------

. logit sta can typ

Logit Estimates                                   Number of obs =      200
                                                  chi2(2)       =    18.14
                                                  Prob > chi2   =   0.0001
Log Likelihood = -91.011956                       Pseudo R2     =   0.0906


---------------------------------------------------------------------------
     sta |     Coef.    Std. Err.      z      P>|z|     [95% Conf. Interval]
---------+-----------------------------------------------------------------
     can |   1.364004    .7817449    1.745    0.081     -.168188    2.896196
     typ |   2.709722    .8598235    3.151    0.002     1.024499    4.394946
   _cons |  -3.820209    .8563559   -4.461    0.000    -5.498636   -2.141782
---------------------------------------------------------------------------
```

```
. logistic sta can typ

Logit Estimates                                    Number of obs =      200
                                                   chi2(2)       =    18.14
                                                   Prob > chi2   =   0.0001
Log Likelihood = -91.011956                        Pseudo R2     =   0.0906

-------------------------------------------------------------------------------
      sta | Odds Ratio   Std. Err.        z      P>|z|      [95% Conf. Interval]
----------+--------------------------------------------------------------------
      can |   3.911825    3.058049     1.745     0.081      .8451949     18.10514
      typ |    15.0251    12.91894     3.151     0.002        2.7857     81.04022
-------------------------------------------------------------------------------
```

Using M-H analysis:

```
. by typ:tabulate sta can

-> typ=          0
          | can
      sta |         0          1 |     Total
----------+----------------------+----------
        0 |        37         14 |        51
        1 |         1          1 |         2
----------+----------------------+----------
    Total |        38         15 |        53


-> typ=          1
          | can
      sta |         0          1 |     Total
----------+----------------------+----------
        0 |       107          2 |       109
        1 |        35          3 |        38
----------+----------------------+----------
    Total |       142          5 |       147
```

$$\hat{OR}_{MH} = \frac{\displaystyle\sum_{i=1}^{2}\frac{a_i d_i}{N_i}}{\displaystyle\sum_{i=1}^{2}\frac{b_i c_i}{N_i}}$$

Evaluating the expression to obtain the M-H estimate of the odds ratio:

i	a_i	b_i	c_i	d_i	N_i	$a_i d_i /N_i$	$b_i c_i /N_i$
1	1	1	14	37	53	0.698	0.264
2	3	35	2	107	147	2.184	0.476
Total						2.882	0.740

$$\hat{OR}_{MH} = \frac{2.882}{0.740} = 3.895$$

The results of these two approaches are quite similar. After controlling for TYP, the odds of dying prior to discharge from the ICU are nearly four times greater in patients with cancer than in patients without cancer.

INTERACTION

Using logistic regression modeling:

```
. gen cantyp=can*typ

. logit sta can typ cantyp

Logit Estimates                                    Number of obs =      200
                                                   chi2(3)       =    18.24
                                                   Prob > chi2   =   0.0004
Log Likelihood = -90.960895                        Pseudo R2     =   0.0911


------------------------------------------------------------------------------
     sta |      Coef.   Std. Err.       z     P>|z|      [95% Conf. Interval]
---------+--------------------------------------------------------------------
     can |   .9718606   1.448604     0.671    0.502     -1.867351    3.811072
     typ |   2.493437   1.03196      2.416    0.016      .4708326    4.516042
  cantyp |   .5510853   1.723283     0.320    0.749     -2.826487    3.928657
   _cons |  -3.610918   1.013422    -3.563    0.000     -5.597189   -1.624647
------------------------------------------------------------------------------
```

The Wald statistic for the interaction term CANTYP indicates that there is no effect modification of the association between CAN and STA by the variable TYP.

Using M-H analysis to test for heterogeneity across strata:

$$\chi_H^2 = \sum_{i=1}^{2} \left\{ w_i \left[\ln\left(\hat{OR}_i\right) - \ln\left(\hat{OR}_L\right) \right]^2 \right\} \qquad \text{where} \quad \hat{OR}_L = \exp\left[\frac{\sum w_i \ln\left(\hat{OR}_i\right)}{\sum w_i} \right]$$

	TYP=0	TYP=1
$\hat{OR}$	2.643	4.586
$\ln(\hat{OR})$	0.972	1.523
$\hat{Var}[\ln(\hat{OR})]$	2.098	0.871
w	0.477	1.148

$$\hat{OR}_L = \exp\left[\frac{0.477(0.972) + 1.148(1.523)}{0.477 + 1.148} \right] = \exp(1.361) = 3.901$$

therefore,

$$\chi_H^2 = \left[(0.477)(0.972 - 1.361)^2 + (1.148)(1.523 - 1.361)^2 \right] = 0.102$$

$$\chi_H^2 \sim \chi^2(1)$$

$$p = 0.749$$

The results of these two approaches are quite similar. There is no indication that TYP is an effect modifier.

2. *Use the data from the Prostate Cancer Study described in Section 1.6.3 to answer the following questions:*

    (a)   *By fitting a series of logistic regression models show that RACE is not a confounder of the PSA CAPSULE odds ratio but is an effect modifier (at the 10 percent level).*

```
. logit capsule psa

Iteration 0:   Log Likelihood =-256.14442
Iteration 1:   Log Likelihood =-233.39251
Iteration 2:   Log Likelihood =-231.64579
Iteration 3:   Log Likelihood =-231.58066
Iteration 4:   Log Likelihood =-231.58055

Logit Estimates                                Number of obs =      380
                                               chi2(1)       =    49.13
                                               Prob > chi2   =   0.0000
Log Likelihood = -231.58055                    Pseudo R2     =   0.0959


------------------------------------------------------------------------------
 capsule |     Coef.    Std. Err.      z      P>|z|     [95% Conf. Interval]
---------+--------------------------------------------------------------------
     psa |   .0501761    .0092502    5.424    0.000      .032046     .0683062
   _cons |  -1.113695    .1615629   -6.893    0.000    -1.430352    -.7970371
------------------------------------------------------------------------------

. logit capsule psa racenew

Iteration 0:   Log Likelihood =-253.80627
Iteration 1:   Log Likelihood =-231.50856
Iteration 2:   Log Likelihood =-229.73691
Iteration 3:   Log Likelihood =-229.67147
Iteration 4:   Log Likelihood =-229.67136

Logit Estimates                                Number of obs =      377
                                               chi2(2)       =    48.27
                                               Prob > chi2   =   0.0000
Log Likelihood = -229.67136                    Pseudo R2     =   0.0951


------------------------------------------------------------------------------
 capsule |     Coef.    Std. Err.      z      P>|z|     [95% Conf. Interval]
---------+--------------------------------------------------------------------
     psa |   .0511505    .0094923    5.389    0.000     .0325459     .069755
 racenew |  -.5788434    .4187205   -1.382    0.167     -1.39952    .2418337
   _cons |  -1.078087    .1622229   -6.646    0.000    -1.396038    -.7601364
------------------------------------------------------------------------------
```

There is a 1.9% increase in the value of the coefficient for PSA when RACE is adjusted for in the model. An increase of this magnitude indicates that RACE does not confound the relationship between PSA and CAPSULE.

```
. gen psaracen=psa*racenew
(3 missing values generated)
```

```
. logit capsule psa racenew psaracen

Iteration 0:  Log Likelihood =-253.80627
Iteration 1:  Log Likelihood =-230.80507
Iteration 2:  Log Likelihood =-228.36276
Iteration 3:  Log Likelihood =-228.21673
Iteration 4:  Log Likelihood =-228.21613
Iteration 5:  Log Likelihood =-228.21613

Logit Estimates                              Number of obs =      377
                                             chi2(3)       =   51.18
                                             Prob > chi2   = 0.0000
Log Likelihood = -228.21613                  Pseudo R2     = 0.1008


------------------------------------------------------------------------------
 capsule |    Coef.   Std. Err.       z     P>|z|    [95% Conf. Interval]
---------+--------------------------------------------------------------------
     psa |  .0607802   .0117133    5.189    0.000     .0378225    .0837379
 racenew |  .0954302   .5420878    0.176    0.860    -.9670424    1.157903
psaracen | -.0349146   .0192711   -1.812    0.070    -.0726852    .0028559
   _cons | -1.190408   .1792926   -6.639    0.000    -1.541815   -.8390007
------------------------------------------------------------------------------
```

When a term representing the interaction between PSA and RACE is added to the model, the term is significant at the $p < 0.10$ level, with $p = 0.070$.

(b)   *Graph the estimated logits from the interactions model versus PSA and interpret the two lines that appear on the graph. Use the graph to illustrate the log-odds of Black vs. White for a subject with PSA=7.0. Use the graph to illustrate the log-odds for a 5-unit increase in PSA for Whites and for Blacks.*

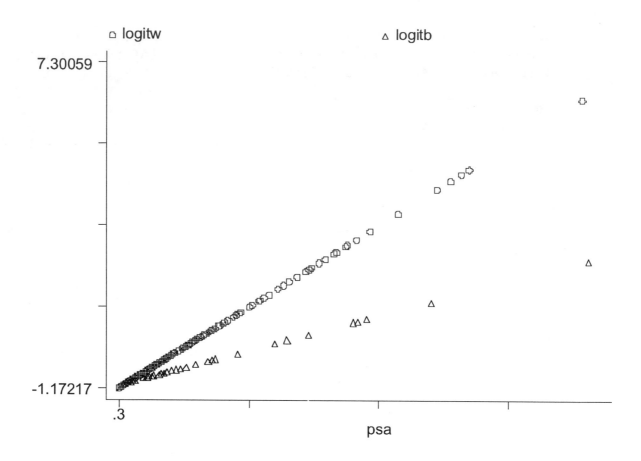

The above graph suggests that the relationship between PSA and CAPSULE differs for White and Black men.   A stronger relationship between PSA and CAPSULE is seen for White men (logitw) than for Black men (logitb).

Use the graph to illustrate the log-odds of Black vs. White for a subject with PSA=7.0.

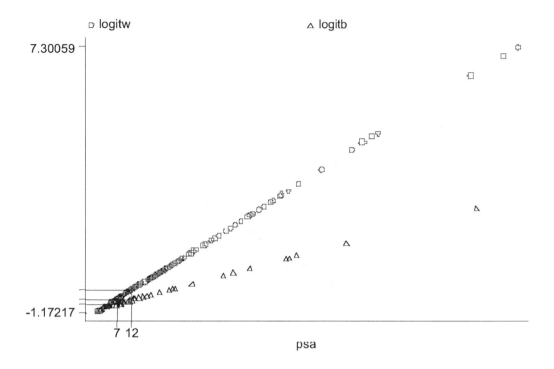

Using the graph above, we see that the logit difference (Log Odds) Black vs. White for a subject with PSA=7.0 is:

$$\ln\left[\hat{OR}(black, white)\right] = -0.9139186 - (-0.7649463)$$

$$= -0.1489723$$

Use the graph to illustrate the log-odds for a 5-unit increase in PSA for Whites and for Blacks.

The graph shown above is enlarged to better show the relation between the logits for blacks and whites among subjects with PSA values less than 15.

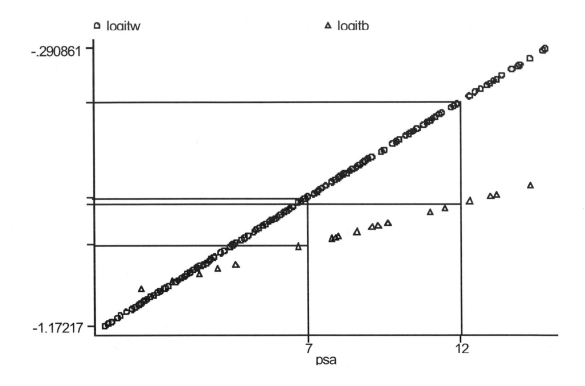

Using the graph above, we see that the logit difference (Log Odds) for a 5-unit increase in PSA level in Black subjects is:

$$\ln\left[\hat{OR}(PSA+5,PSA)\right] = -0.7845908 - (-0.9139186)$$

$$= 0.1293278$$

Using the graph above, we see that the logit difference (Log Odds) for a 5-unit increase in PSA level in White subjects is:

$$\ln\left[\hat{OR}(PSA+5,PSA)\right] = -0.4610453 - (-0.7649463)$$

$$= 0.303901$$

The choice of PSA levels of 7 and 12 for illustration is arbitrary. The relationship between PSA and CAPSULE within each race group is constant.

(c) *Estimate the point and 95 percent confidence interval estimates of the odds ratios corresponding to each of the log-odds illustrated in Exercise 2(b). Add the 95% confidence bands to the graph of the estimated logits from the interactions model in Exercise 2(b). Transform the lines and bands in this plot to obtain a plot of the estimated probability with its 95% confidence bands. Use the graph to estimate, point and interval, the probability of penetration for both a White and Black with PSA=7.0. Interpret the two point and interval estimates.*

To estimate confidence intervals, obtain the covariance matrix.

```
. corr, _coef covariance

        |       psa    racenew psaracen      _cons
--------+-------------------------------------------
    psa| .000137
 racenew| .001577   .293859
psaracen| -.000137 -.007142    .000371
   _cons| -.001577 -.032146    .001577    .032146
```

The logit difference (Log Odds) Black vs. White for a subject with PSA = 7.0 is:

$$\ln\left[\hat{OR}(black, white)\right] = -0.9139186 - (-0.7649463) = -0.1489723$$

therefore,

$$\hat{OR} = 0.86$$

To calculate the confidence interval for the odds ratio, calculate the standard error:

$$Var\left[\ln\left[\hat{OR}(black, white)\right]\right] = Var\left(\hat{\beta}_2\right) + x^2 Var\left(\hat{\beta}_3\right) + 2xCov\left(\hat{\beta}_2, \hat{\beta}_3\right)$$

$$= (0.293859) + (7)^2(0.000371) + 2(7)(-0.007142)$$

$$= 0.21205$$

$$\hat{SE}\left[\ln\left[\hat{OR}(black, white)\right]\right] = sqrt(0.21205) = 0.4605$$

$$95\%CI\left[\ln\left[\hat{OR}(black, white)\right]\right] = -0.1489723 \pm 1.96(0.4605)$$

$$-1.0515 \leq \ln\left[\hat{OR}(black, white)\right] \leq 0.75359$$

therefore,

$$0.35 \leq \hat{OR}(black, white) \leq 2.12$$

This means that a Black subject with a PSA of 7.0 is 0.86 times as likely to have tumor penetration of the prostate capsule than a white subject with the same PSA. The confidence interval suggests that this difference could be as little as 0.35 times or as much as 2.12 times with 95% confidence.

The logit difference (Log Odds) for a 5-unit increase in PSA level in Black subjects is:

BLACK SUBJECTS:

$$\ln\left[\hat{OR}(PSA+5, PSA)\right] = -0.7845908 - (-0.9139186) = 0.1293278$$

therefore,

$$\hat{OR} = 1.14$$

To calculate the confidence interval for the odds ratio, calculate the standard error:

BLACK SUBJECTS:

$$\hat{Var}\left[\ln\left[\hat{OR}(PSA+5, PSA)\right]\right] = x^2 \hat{Var}\left(\hat{\beta}_1\right) + x^2 \hat{Var}\left(\hat{\beta}_3\right) + 2x^2 \hat{Cov}\left(\hat{\beta}_1, \hat{\beta}_3\right)$$

$$= (5)^2(0.000137) + (5)^2(0.000371) + 2(5)(5)(-0.000137)$$

$$= 0.00585$$

$$\hat{SE}\left[\ln\left[\hat{OR}(PSA+5, PSA)\right]\right] = sqrt(0.00585) = 0.0765$$

$$95\%CI\left[\ln\left[\hat{OR}(PSA+5, PSA)\right]\right] = 0.1293278 \pm 1.96(0.0765)$$

$$-0.02058 \le \ln\left[\hat{OR}(PSA+5, PSA)\right] \le 0.27924$$

therefore,

$$0.98 \le OR(PSA+5, PSA) \le 1.32$$

This means that among Black subjects, a 5-unit increase in PSA level is associated with 1.14 times the risk of tumor penetration of the prostate capsule. The confidence intervals suggests that the odds ratio could be as low as 0.98 or as high as 1.32 with 95% confidence.

The logit difference (Log Odds) for a 5-unit increase in PSA level in White subjects is:

WHITE SUBJECTS:

$$\ln\left[\hat{OR}(PSA+5,PSA)\right] = -0.4610453 - (-0.7649463) = 0.3039$$

therefore,

$$\hat{OR} = 1.36$$

To calculate the confidence interval for the odds ratio, calculate the standard error:

WHITE SUBJECTS:

$$\hat{Var}\left[\ln\left[\hat{OR}(PSA+5,PSA)\right]\right] = x^2\hat{Var}\left(\hat{\beta}_1\right)$$

$$= (5)^2(0.000137) = 0.003425$$

$$\hat{SE}\left[\ln\left[\hat{OR}(PSA+5,PSA)\right]\right] = sqrt(0.003425) = 0.0585$$

$$95\%CI\left[\ln\left[\hat{OR}(PSA+5,PSA)\right]\right] = 0.3039 + 1.96(0.0585)$$

$$0.18919 \le \ln\left[\hat{OR}(PSA+5,PSA)\right] \le 0.41861$$

therefore,

$$1.21 \le OR(PSA+5,PSA) \le 1.52$$

This means that among White subjects, a 5-unit increase in PSA level is associated with 1.36 times the risk of tumor penetration of the prostate capsule. The confidence intervals suggests that the odds ratio could be as low as 1.21 or as high as 1.52 with 95% confidence.

Add the 95% confidence bands to the graph of the estimated logits from the interactions model in 2(b).

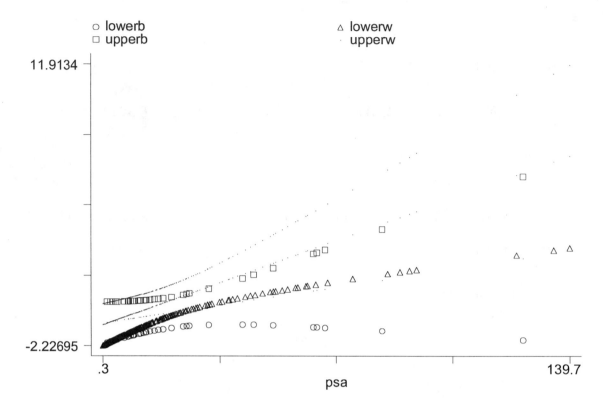

Transform the lines and bands in this plot to obtain a plot of the estimated probability with its 95% confidence bands.

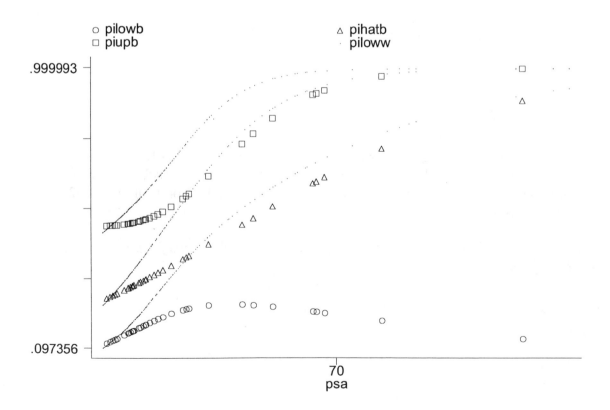

Use the graph to estimate, point and interval, the probability of penetration for both a White and Black with PSA=7.0. Interpret the two point and interval estimates.

The probability of capsule penetration for a Black subject with PSA=7.0 is 28.6%. This means that of a population of Black subjects with this PSA value, 28.6% would have capsule penetration. This value could be as low as 14.1% or as high as 49.3% with 95% confidence.

The probability of capsule penetration for a White subject with PSA=7.0 is 31.8%. This means that of a population of White subjects with this PSA value, 31.8% would have capsule penetration. This value could be as low as 15.8% or as high as 53.4% with 95% confidence.

Chapter Four – Solutions

1. *Selection of the scale for continuous covariates is an important step in any modeling process. The variable systolic blood pressure at admission, SYS, in the ICU study described in Section 1.6 presents a particularly challenging example. Consider the variable vital status (STA) as the outcome variable and SYS as the covariate for a univariable logistic regression model. What is the correct scale for SYS to enter the model? As a second example, consider a univariable model with heart rate at ICU admission (HRA) as the covariate. Repeat this exercise of scale identification for SYS and HRA using multivariable model containing these two variables plus three or four other covariates of your choice.*

The variable systolic blood pressure at admission, SYS, in the ICU study described in Section 1.6 presents a particularly challenging example. Consider the variable vital status (STA) as the outcome variable and SYS as the covariate for a univariable logistic regression model. What is the correct scale for SYS to enter the model?

Use quartile design variables:

```
. sum sys, det

                             sys
-------------------------------------------------------------
      Percentiles      Smallest
 1%          52              36
 5%          80              48
10%          92              56        Obs            200
25%         110              62        Sum of Wgt.    200

50%         130                        Mean        132.28
                         Largest       Std. Dev.   32.9521
75%         150             208
90%         170             212        Variance    1085.841
95%         190             224        Skewness    .2965964
99%         218             256        Kurtosis    4.028003

. tab sysgp

     sysgp |      Freq.     Percent        Cum.
-----------+-----------------------------------
         1 |         53       26.50       26.50
         2 |         48       24.00       50.50
         3 |         52       26.00       76.50
         4 |         47       23.50      100.00
-----------+-----------------------------------
     Total |        200      100.00
```

```
. logit STA sys_2 sys_3 sys_4

Iteration 0:   Log Likelihood =-100.08048
Iteration 1:   Log Likelihood =-96.324305
Iteration 2:   Log Likelihood =-96.136717
Iteration 3:   Log Likelihood =-96.135338
Iteration 4:   Log Likelihood =-96.135338

Logit Estimates                               Number of obs =      200
                                              chi2(3)       =     7.89
                                              Prob > chi2   = 0.0483
Log Likelihood = -96.135338                   Pseudo R2     = 0.0394

---------------------------------------------------------------------
     STA |     Coef.   Std. Err.        z    P>|z|    [ 95% Conf. Interval]
---------+-----------------------------------------------------------
   sys_2 | -.6280079   .4756881    -1.320    0.187    -1.560339    .3043237
   sys_3 | -.4773476   .4525775    -1.055    0.292    -1.364383    .4096879
   sys_4 | -1.536577   .6023146    -2.551    0.011    -2.717091   -.3560617
    _cons | -.8383292   .2992107    -2.802    0.005    -1.424771    -.251887
---------------------------------------------------------------------
```

Plot the estimated logistic regression coefficients versus the quartile midpoints:

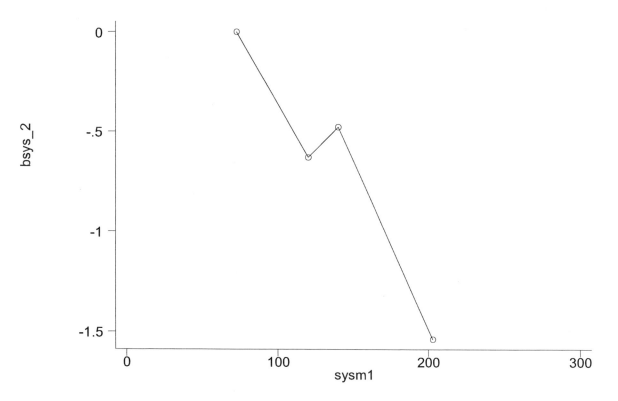

We see that there is tendency toward a linear decrease; however, the log-odds of the third quartile vs. the first quartile is a bit out of line.

Use a scatterplot smooth:

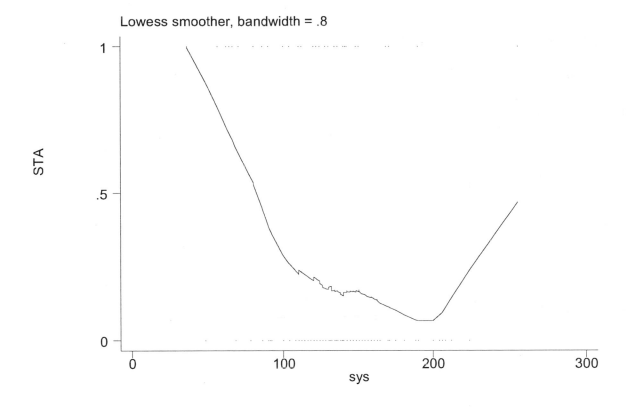

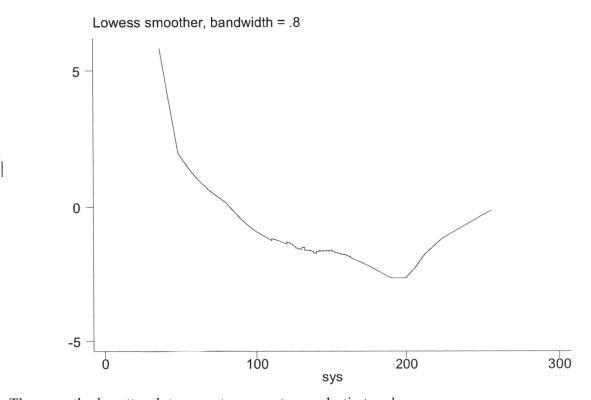

The smoothed scatterplot seems to suggest a quadratic trend.

Use the method of fractional polynomials:

```
. fracpoly logit STA sys, compare
........
-> gen sys_1 = x^.5 if _sample
-> gen sys_2 = x^3 if _sample
where: x = sys/100

Iteration 0:   Log Likelihood =-100.08048
Iteration 1:   Log Likelihood =-92.255518
Iteration 2:   Log Likelihood =-91.786945
Iteration 3:   Log Likelihood =-91.783861
Iteration 4:   Log Likelihood =-91.783861

Logit Estimates                         Number of obs =     200
                                        chi2(2)       =   16.59
                                        Prob > chi2   =  0.0002
Log Likelihood = -91.783861             Pseudo R2     =  0.0829


------------------------------------------------------------------------
     STA |      Coef.   Std. Err.       z     P>|z|    [ 95% Conf. Interval]
---------+--------------------------------------------------------------
   sys_1 |  -9.014211   2.500338    -3.605    0.000    -13.91478   -4.113639
   sys_2 |   .4259721   .1659321     2.567    0.010     .1007512    .7511929
   _cons |   7.627334   2.433821     3.134    0.002     2.857134    12.39754
------------------------------------------------------------------------
Deviance:  183.568. Best powers of sys among 44 models fit: .5 3.

Fractional Polynomial Model Comparisons:
------------------------------------------------------------------------
sys             df       Deviance    Gain    P(term)  Powers
------------------------------------------------------------------------
Not in model     0        200.161     --       --
Linear           1        191.335    0.000    0.003    1
m = 1            2        186.349    4.986    0.026    -1
m = 2            4        183.568    7.767    0.249    .5 3
------------------------------------------------------------------------
```

The fractional polynomial analysis suggests that using the inverse (1/sys) is a significant improvement over using sys.

```
. gen invsys=1/(sys)

. logit STA invsys

Iteration 0:   Log Likelihood =-100.08048
Iteration 1:   Log Likelihood =-93.353131
Iteration 2:   Log Likelihood =-93.174523
Iteration 3:   Log Likelihood =-93.174348

Logit Estimates                         Number of obs =     200
                                        chi2(1)       =   13.81
                                        Prob > chi2   =  0.0002
Log Likelihood = -93.174348             Pseudo R2     =  0.0690


------------------------------------------------------------------------
     STA |      Coef.   Std. Err.       z     P>|z|    [ 95% Conf. Interval]
---------+--------------------------------------------------------------
  invsys |   236.0774   71.17957     3.317    0.001    96.56805    375.5868
   _cons |  -3.384721   .6391614    -5.296    0.000    -4.637455   -2.131988
------------------------------------------------------------------------
```

As a second example, consider a univariable model with heart rate at ICU admission (HRA) as the covariate.

```
. sum hra, det

                              hra
-------------------------------------------------------------
        Percentiles      Smallest
  1%          45              39
  5%          60              44
 10%          65              46        Obs                 200
 25%          80              48        Sum of Wgt.         200

 50%          96                        Mean             98.925
                          Largest       Std. Dev.      26.82962
 75%         118.5          160
 90%         136.5          162         Variance       719.8285
 95%         144.5          170         Skewness       .4115287
 99%         166            192         Kurtosis       3.044908

. tab hragp

      hragp |      Freq.      Percent        Cum.
------------+-----------------------------------
          1 |         52        26.00       26.00
          2 |         53        26.50       52.50
          3 |         45        22.50       75.00
          4 |         50        25.00      100.00
------------+-----------------------------------
      Total |        200       100.00
```

```
. logit sta hra_2 hra_3 hra_4

Iteration 0:   Log Likelihood =-100.08048
Iteration 1:   Log Likelihood =-99.714701
Iteration 2:   Log Likelihood =-99.713899
Iteration 3:   Log Likelihood =-99.713899

Logit Estimates                           Number of obs =      200
                                          chi2(3)       =     0.73
                                          Prob > chi2   = 0.8654
Log Likelihood = -99.713899               Pseudo R2     = 0.0037

------------------------------------------------------------------------------
     sta |      Coef.    Std. Err.       z      P>|z|     [95% Conf. Interval]
---------+--------------------------------------------------------------------
   hra_2 |    .3353101    .4920269     0.681    0.496    -.6290449    1.299665
   hra_3 |    .0324992    .5351579     0.061    0.952    -1.016391    1.081389
   hra_4 |    .2983092    .5009162     0.596    0.551    -.6834686    1.280087
   _cons |   -1.563976    .3665609    -4.267    0.000    -2.282422   -.8455293
------------------------------------------------------------------------------
```

Plot the estimated logistic regression coefficients versus the quartile midpoints:

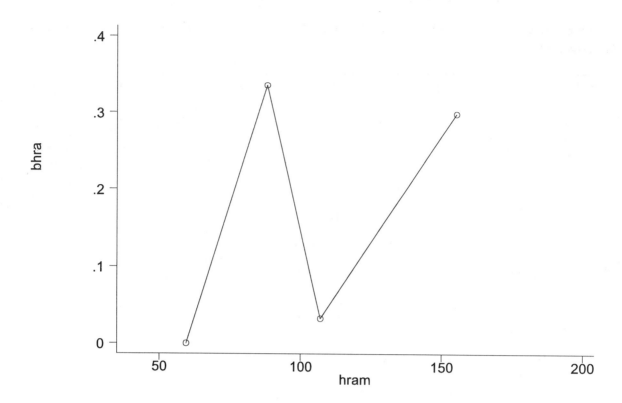

We see that there does not appear to be a relationship between STA and HRA.

Use a scatterplot smooth:

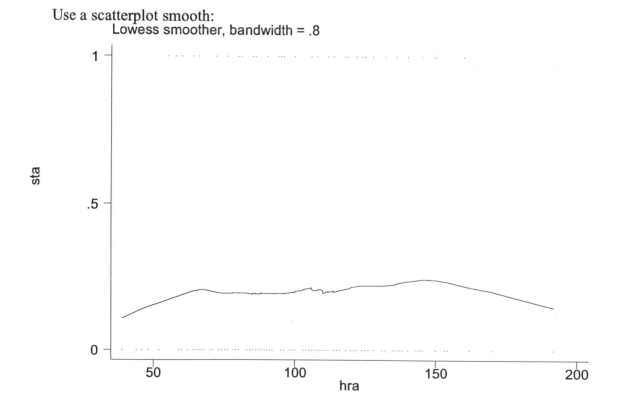

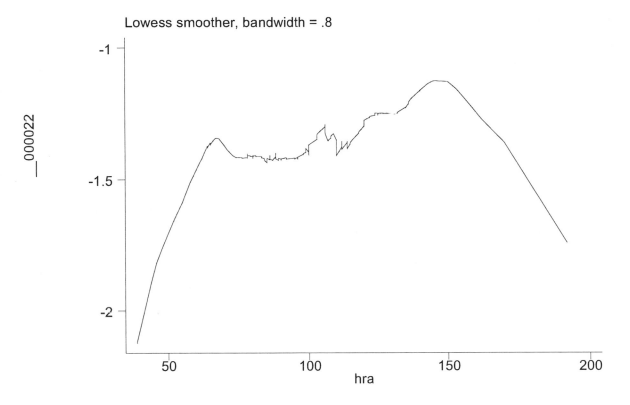

The smoothed scatterplot confirms that there does not seem to be a strong relationship between STA and HRA.

Use the method of fractional polynomials:

```
. fracpoly logit sta hra, compare
........
-> gen hra_1 = x^-2 if _sample
-> gen hra_2 = x^-2*ln(x) if _sample
where: x = hra/100

Iteration 0:   Log Likelihood =-100.08048
Iteration 1:   Log Likelihood =-99.818535
Iteration 2:   Log Likelihood =-99.803522
Iteration 3:   Log Likelihood =-99.803404
Iteration 4:   Log Likelihood =-99.803404

Logit Estimates                              Number of obs =      200
                                             chi2(2)       =     0.55
                                             Prob > chi2   =   0.7580
Log Likelihood = -99.803404                  Pseudo R2     =   0.0028


------------------------------------------------------------------------------
     sta |      Coef.   Std. Err.       z     P>|z|     [95% Conf. Interval]
---------+--------------------------------------------------------------------
   hra_1 |   .2061483   .9577467     0.215    0.830    -1.671001     2.083297
   hra_2 |   .4307214   1.180929     0.365    0.715    -1.883857       2.7453
   _cons |  -1.538297   .9465618    -1.625    0.104    -3.393524     .3169302
------------------------------------------------------------------------------
Deviance:  199.607. Best powers of hra among 44 models fit: -2 -2.

Fractional Polynomial Model Comparisons:
------------------------------------------------------------------
hra             df      Deviance    Gain   P(term) Powers
------------------------------------------------------------------
Not in model     0       200.161     --      --
Linear           1       199.960   0.000   0.654   1
m = 1            2       199.750   0.211   0.646   -2
m = 2            4       199.607   0.354   0.931   -2 -2
------------------------------------------------------------------
```

The fractional polynomial analysis also suggests that there is not a significant relationship between STA and HRA. If there is clinical evidence to support the inclusion of HRA in models of STA, it appears that it is appropriate to include this variable in its linear form.

Repeat this exercise of scale identification for SYS and HRA using a multivariable model containing these two variables plus three or four other covariates of your choice.

Consider the regression of vital status, STA, on systolic blood pressure (STA), heart rate (HRA), age, cancer (CAN), type of admission (TYP) and level of consciousness (LOC). For the purpose of this set of analyses only, the variable LOC has been recoded as a dichotomous variable LOC2 with 0 = No coma or deep stupor and 1 = Coma or deep stupor.

```
. logit sta sys hra age can typ loc2

Iteration 0:  Log Likelihood =-100.08048
Iteration 1:  Log Likelihood =-70.921109
Iteration 2:  Log Likelihood =-68.091626
Iteration 3:  Log Likelihood =-67.811173
Iteration 4:  Log Likelihood =-67.804813
Iteration 5:  Log Likelihood =-67.804808

Logit Estimates                          Number of obs =      200
                                         chi2(6)       =    64.55
                                         Prob > chi2   =   0.0000
Log Likelihood = -67.804808              Pseudo R2     =   0.3225

------------------------------------------------------------------------
    sta |     Coef.    Std. Err.      z     P>|z|     [95% Conf. Interval]
--------+---------------------------------------------------------------
    sys |  -.0125144    .0069625    -1.797   0.072    -.0261607    .0011319
    hra |  -.0003711    .008474     -0.044   0.965    -.0169799    .0162377
    age |   .0366338    .0128331     2.855   0.004     .0114813    .0617863
    can |   2.138251    .8439735     2.534   0.011     .4840935    3.792409
    typ |   2.965739    .9228027     3.214   0.001     1.157079    4.774399
   loc2 |   3.848226    .9572424     4.020   0.000     1.972065    5.724386
   _cons |  -5.071054    1.836741    -2.761   0.006    -8.670999   -1.471108
------------------------------------------------------------------------
```

Use the method of fractional polynomials to evaluate the scale of the continuous covariates in the multivariable model:

```
. fracpoly logit sta sys hra age can typ loc2, compare
........
-> gen sys_1 = x^-2 if _sample
-> gen sys_2 = x^-1 if _sample
where: x = sys/100

Iteration 0:  Log Likelihood =-100.08048
Iteration 1:  Log Likelihood =-70.732456
Iteration 2:  Log Likelihood =-67.913573
Iteration 3:  Log Likelihood =-67.614339
Iteration 4:  Log Likelihood =-67.607612
Iteration 5:  Log Likelihood =-67.607607

Logit Estimates                         Number of obs =      200
                                        chi2(7)       =    64.95
                                        Prob > chi2   = 0.0000
Log Likelihood = -67.607607             Pseudo R2     = 0.3245


------------------------------------------------------------------------
    sta |     Coef.   Std. Err.       z    P>|z|    [95% Conf. Interval]
--------+---------------------------------------------------------------
  sys_1 |  -1.31476   1.111063    -1.183   0.237   -3.492404    .8628836
  sys_2 |  4.422753   2.850066     1.552   0.121   -1.163273   10.00878
    hra |  -.0010199    .008544    -0.119   0.905   -.0177657    .015726
    age |   .0363166   .012903     2.815   0.005    .0110271   .0616061
    can |   2.128522   .8491284    2.507   0.012    .4642612   3.792784
    typ |   2.956773   .9263498    3.192   0.001    1.141161   4.772385
   loc2 |   3.852801   .9968108    3.865   0.000    1.899088   5.806514
  _cons |  -9.277982   2.111651    -4.394   0.000   -13.41674  -5.139221
------------------------------------------------------------------------
Deviance:  135.215. Best powers of sys among 44 models fit: -2 -1.

Fractional Polynomial Model Comparisons:
-------------------------------------------------------------
sys             df     Deviance     Gain    P(term) Powers
-------------------------------------------------------------
Not in model     0     139.044       --        --
Linear           1     135.610     0.000     0.064  1
m = 1            2     135.584     0.025     0.874  .5
m = 2            4     135.215     0.394     0.831  -2 -1
```

It does not appear that there is any advantage to transforming SYS, in fact, it does not appear that the inclusion of SYS improves the model.

```
. fracpoly logit sta hra sys age can typ loc2, compare
........
-> gen hra_1 = x^-2 if _sample
-> gen hra_2 = x^-2*ln(x) if _sample
where: x = hra/100

Iteration 0:   Log Likelihood =-100.08048
Iteration 1:   Log Likelihood =-67.463621
Iteration 2:   Log Likelihood =-64.037451
Iteration 3:   Log Likelihood =-63.591273
Iteration 4:   Log Likelihood =-63.574115
Iteration 5:   Log Likelihood =-63.574074

Logit Estimates                          Number of obs =      200
                                         chi2(7)       =    73.01
                                         Prob > chi2   =   0.0000
Log Likelihood = -63.574074              Pseudo R2     =   0.3648


------------------------------------------------------------------------
    sta |     Coef.   Std. Err.      z     P>|z|     [95% Conf. Interval]
--------+---------------------------------------------------------------
  hra_1 |   2.464465   1.247407    1.976   0.048     .0195911    4.909339
  hra_2 |   3.390792   1.499867    2.261   0.024      .451107    6.330477
    sys |  -.0195088   .0079696   -2.448   0.014    -.0351289   -.0038887
    age |   .0388751   .0133087    2.921   0.003     .0127905    .0649597
    can |   2.169674   .8692556    2.496   0.013     .4659645    3.873384
    typ |   3.113619   .9722872    3.202   0.001     1.207971    5.019266
   loc2 |    5.24225   1.347221    3.891   0.000     2.601744    7.882755
  _cons |  -6.829938   1.956807   -3.490   0.000    -10.66521   -2.994667
------------------------------------------------------------------------
Deviance:   127.148. Best powers of hra among 44 models fit: -2 -2.

Fractional Polynomial Model Comparisons:
-----------------------------------------------------------------
hra               df      Deviance    Gain   P(term) Powers
-----------------------------------------------------------------
Not in model      0       135.612      --      --
Linear            1       135.610     0.000   0.965   1
m = 1             2       133.911     1.699   0.192   -2
m = 2             4       127.148     8.461   0.034   -2 -2
-----------------------------------------------------------------
```

Using the -2 -2 transformation of HRA, significantly improves the model compared with using HRA in its linear form. Using the fracplot option (output not shown) supports this claim. Does this change in HRA influence the scale for SYS?

```
. fracpoly logit sta sys hra_1 hra_2 age can typ loc2, compare
........
-> gen sys_1 = x^-2 if _sample
-> gen sys_2 = x^3 if _sample
where: x = sys/100

Iteration 0:  Log Likelihood =-100.08048
Iteration 1:  Log Likelihood =-67.415885
Iteration 2:  Log Likelihood =-63.173484
Iteration 3:  Log Likelihood =-62.697119
Iteration 4:  Log Likelihood =-62.680208
Iteration 5:  Log Likelihood =-62.680168


Logit Estimates                             Number of obs  =      200
                                            chi2(8)        =    74.80
                                            Prob > chi2    =   0.0000
Log Likelihood = -62.680168                 Pseudo R2      =   0.3737


-------------------------------------------------------------------------
    sta |    Coef.    Std. Err.      z     P>|z|     [95% Conf. Interval]
--------+----------------------------------------------------------------
  sys_1 |  1.151272   .6007918     1.916   0.055    -.0262581    2.328802
  sys_2 | -.0787732   .1443086    -0.546   0.585    -.3616129    .2040664
  hra_1 |  3.175471   1.355622     2.342   0.019     .5185001    5.832441
  hra_2 |  4.402831   1.644648     2.677   0.007     1.17938     7.626282
    age |  .0373453   .0135451     2.757   0.006     .0107974    .0638931
    can |  2.097335   .8704398     2.410   0.016     .3913048    3.803366
    typ |    3.0249   .9649301     3.135   0.002     1.133672    4.916128
   loc2 |  5.042497   1.339176     3.765   0.000     2.41776     7.667233
  _cons | -10.47409   2.236942    -4.682   0.000    -14.85842   -6.089764
-------------------------------------------------------------------------
Deviance:  125.360. Best powers of sys among 44 models fit: -2 3.

Fractional Polynomial Model Comparisons:
--------------------------------------------------------------
sys           df      Deviance    Gain    P(term)  Powers
--------------------------------------------------------------
Not in model   0      133.784      --       --
Linear         1      127.148    0.000    0.010   1
m = 1          2      125.652    1.496    0.221   -2
m = 2          4      125.360    1.788    0.864   -2 3
--------------------------------------------------------------
```

Use the −2 −2 transformation for HRA, keep SYS in its linear form. Age is included in the model is continuous, use the method of fractional polynomials to check the appropriate scale for the variable AGE.

```
. fracpoly logit sta age hra_1 hra_2 sys can typ loc2, compare
........
-> gen age_1 = x^-.5 if _sample
-> gen age_2 = ln(x) if _sample
where: x = age/10

Iteration 0:  Log Likelihood =-100.08048
Iteration 1:  Log Likelihood =-67.141377
Iteration 2:  Log Likelihood =-63.718743
Iteration 3:  Log Likelihood =-63.248748
Iteration 4:  Log Likelihood =-63.229286
Iteration 5:  Log Likelihood =-63.229233

Logit Estimates                          Number of obs =      200
                                         chi2(8)       =    73.70
                                         Prob > chi2   =   0.0000
Log Likelihood = -63.229233              Pseudo R2     =   0.3682


------------------------------------------------------------------------
     sta |     Coef.   Std. Err.       z    P>|z|    [95% Conf. Interval]
---------+--------------------------------------------------------------
   age_1 |  28.99209   19.34394    1.499    0.134   -8.921341    66.90552
   age_2 |  8.885633   4.908322    1.810    0.070   -.7345021    18.50577
   hra_1 |  2.486731   1.250058    1.989    0.047    .0366612      4.9368
   hra_2 |   3.41606   1.503537    2.272    0.023    .4691826    6.362937
     sys | -.0196153   .0080526   -2.436    0.015   -.0353982   -.0038325
     can |   2.30263   .8924504    2.580    0.010    .5534591      4.0518
     typ |  3.186642   .9879148    3.226    0.001    1.250365     5.12292
    loc2 |   5.38741   1.370013    3.932    0.000    2.702235    8.072586
   _cons | -32.44759   17.08087   -1.900    0.057   -65.92548    1.030292
------------------------------------------------------------------------
Deviance:  126.458. Best powers of age among 44 models fit: -.5 0.

Fractional Polynomial Model Comparisons:
----------------------------------------------------------------
age             df     Deviance    Gain   P(term) Powers
----------------------------------------------------------------
Not in model     0      137.529     --       --
Linear           1      127.148    0.000    0.001   1
m = 1            2      126.835    0.313    0.576   2
m = 2            4      126.458    0.690    0.828   -.5 0
----------------------------------------------------------------
```

Include AGE in its linear continuous form. The final model is:

```
. logit sta  hra_1 hra_2 sys age can typ loc2

Iteration 0:  Log Likelihood =-100.08048
Iteration 1:  Log Likelihood =-67.463621
Iteration 2:  Log Likelihood =-64.037451
Iteration 3:  Log Likelihood =-63.591273
Iteration 4:  Log Likelihood =-63.574115
Iteration 5:  Log Likelihood =-63.574074

Logit Estimates                          Number of obs =      200
                                         chi2(7)       =    73.01
                                         Prob > chi2   =   0.0000
Log Likelihood = -63.574074              Pseudo R2     =   0.3648


------------------------------------------------------------------------
     sta |     Coef.   Std. Err.       z    P>|z|    [95% Conf. Interval]
---------+--------------------------------------------------------------
   hra_1 |  2.464465   1.247407    1.976    0.048    .0195911    4.909339
   hra_2 |  3.390792   1.499867    2.261    0.024     .451107    6.330477
     sys | -.0195088   .0079696   -2.448    0.014   -.0351289   -.0038887
     age |  .0388751   .0133087    2.921    0.003    .0127905    .0649597
     can |  2.169674   .8692556    2.496    0.013    .4659645    3.873384
     typ |  3.113619   .9722872    3.202    0.001    1.207971    5.019266
    loc2 |   5.24225   1.347221    3.891    0.000    2.601744    7.882755
   _cons | -6.829938   1.956807   -3.490    0.000   -10.66521   -2.994667
```

2. *Consider the variable level of consciousness at ICU admission (LOC) as a covariate and vital status (STA) as the outcome variable. Compare the estimates of the odds ratios obtained from the cross-classification of STA by LOC and the logistic regression of STA on LOC. Use LOC=0 as the reference group for both methods. How well did the logistic regression program deal with the zero cell? What strategy would you adopt to model LOC in future analyses?*

Cross-classification of STA by LOC:

```
. tab loc sta, row col chi2

           | sta
       loc |         0          1 |     Total
-----------+----------------------+----------
         0 |       158         27 |       185
           |     85.41      14.59 |    100.00
           |     98.75      67.50 |     92.50
-----------+----------------------+----------
         1 |         0          5 |         5
           |      0.00     100.00 |    100.00
           |      0.00      12.50 |      2.50
-----------+----------------------+----------
         2 |         2          8 |        10
           |     20.00      80.00 |    100.00
           |      1.25      20.00 |      5.00
-----------+----------------------+----------
     Total |       160         40 |       200
           |     80.00      20.00 |    100.00
           |    100.00     100.00 |    100.00

       Pearson chi2(2)  =   45.8784   Pr = 0.000
```

Compute Odds Ratios Comparing LOC=1 with LOC=0

```
. drop if loc==2
(10 observations deleted)

. cc sta loc_1

                              loc_1
                                                      Proportion
                 | Exposed   Unexposed |    Total       Exposed
-----------------+---------------------+------------------------
          Cases  |       5          27 |       32        0.1562
        Controls |       0         158 |      158        0.0000
-----------------+---------------------+------------------------
          Total  |       5         185 |      190        0.0263
                 |                     |
                 |    Pt. Est.         | [95% Conf. Interval]
                 |---------------------+------------------------
      Odds ratio |           .         | 7.316192         .   (Cornfield)
   Attr. frac. ex.|          1         | .8633169         .   (Cornfield)
   Attr. frac. pop|      .15625        |
                 +---------------------------------------------
                       chi2(1)  =     25.35  Pr>chi2 = 0.0000
```

The odds ratio is undefined.

Compute Odds Ratios Comparing LOC=2 with LOC=0

```
. *clear the previous selection, begin with full data set

. drop if loc==1

(5 observations deleted)

. cc sta loc_2

                              loc_2
                                                    Proportion
                |  Exposed   Unexposed  |   Total     Exposed
----------------+-----------------------+----------------------
         Cases  |     8           27    |     35       0.2286
       Controls |     2          158    |    160       0.0125
----------------+-----------------------+----------------------
         Total  |    10          185    |    195       0.0513
                |                        |
                |  Pt. Est.              |  [95% Conf. Interval]
                |------------------------+----------------------
    Odds ratio  |   23.40741            |   5.26946           .   (Cornfield)
 Attr. frac. ex.|    .9572785           |   .8102272          .   (Cornfield)
 Attr. frac. pop|    .2188065           |
                +------------------------------------------------
                         chi2(1) =    27.50  Pr>chi2 = 0.0000
```

The odds ratio is 23.41.

Logistic Regression of STA on LOC:

Generate indicator variables for the levels of LOC.

```
. gen loc_1=1 if loc==1
(195 missing values generated)

. replace loc_1=0 if loc_1==.
(195 real changes made)

. gen loc_2=1 if loc==2
(190 missing values generated)

. replace loc_2=0 if loc_2==.
(190 real changes made)
```

```
. logistic sta loc_1 loc_2

Note: loc_1~=0 predicts success perfectly
      loc_1 dropped and 5 obs not used


Logit Estimates                            Number of obs =      195
                                           chi2(1)       =    19.76
                                           Prob > chi2   =   0.0000
Log Likelihood = -81.892241                Pseudo R2     =   0.1076


------------------------------------------------------------------------
     sta | Odds Ratio   Std. Err.      z    P>|z|   [95% Conf. Interval]
---------+--------------------------------------------------------------
   loc_2 |   23.40741   19.13635     3.857   0.000   4.714842   116.2089
------------------------------------------------------------------------
```

The odds ratio is 23.41.

STATA tells the user that when LOC=1, the outcome STA is predicted perfectly. The program drops observations with LOC=1 and performs the regression on the remaining subjects.

The cross-classification of STA by LOC suggests that patients in either deep stupor or coma are more likely to die than patients with no deep stupor or coma. This suggests that a good strategy for modeling LOC in future analyses is to create a dichotomous covariate LOC2, which takes on the value 0 if the patient had neither deep stupor nor coma, and the value 1 if the patient had either deep stupor or coma. The cross-classification of the original variable LOC by the new variable LOC2 is shown below.

```
. tab loc loc2

          | loc2
     loc |          0          1 |      Total
---------+----------------------+----------
       0 |        185          0 |        185
       1 |          0          5 |          5
       2 |          0         10 |         10
---------+----------------------+----------
   Total |        185         15 |        200
```

3. *Consider the variable vital status (STA) as the outcome variable and the remainder of the variables in the ICU data set as potential covariates. Use each of the variable selection methods discussed in this chapter to find a "best" model. Document thoroughly the rationale for each step in each process you follow. Compare and contrast the models resulting from the different approaches to variable selection. Note that in all cases the analysis should address not only identification of main effects but also appropriate scale for continuous covariates and potential interactions. Display the results of your final model in a table. Include in the table point and 95% CI estimates of all relevant odds ratios. Document the rationale for choosing the final model.*

Purposeful Selection:

Begin with a univariate description of all covariates. An example of this analysis for a catgorical and a continuous variable is shown below.

```
. tab sex

        sex |      Freq.      Percent        Cum.
------------+-----------------------------------
          0 |        124        62.00       62.00
          1 |         76        38.00      100.00
------------+-----------------------------------
      Total |        200       100.00

. sum hra, det

                            hra
-------------------------------------------------------------
      Percentiles      Smallest
 1%           45            39
 5%           60            44
10%           65            46         Obs                 200
25%           80            48         Sum of Wgt.         200

50%           96                       Mean            98.925
                         Largest       Std. Dev.     26.82962
75%        118.5            160
90%        136.5            162         Variance      719.8285
95%        144.5            170         Skewness     .4115287
99%          166            192         Kurtosis     3.044908
```

The univariate analysis does not reveal any variables for which there are illegal values. For each of the catcgorical variables, at least 1% of cases has a value greater than 0. All binary variables are coded as 0/1. Race (RACE) and Level of Consciousness at ICU Admission (LOC) are the only non-binary categorical variables. Indicator variables for RACE are created.

RACE	Label	RACE_2	RACE_3
1	White	0	0
2	Black	1	0
3	Other	0	1

As discussed in the previous problem, a binary variable for level of consciousness is created, LOC2. This variable takes on the value 0 if the patient had neither deep stupor nor coma, and the value 1 if the patient had either deep stupor or coma. The cross-classification of the original variable LOC by the new variable LOC2 is shown below.

```
. tab loc loc2

            | loc2
        loc |         0         1 |     Total
------------+----------------------+----------
          0 |       185         0 |       185
          1 |         0         5 |         5
          2 |         0        10 |        10
------------+----------------------+----------
      Total |       185        15 |       200
```

The three continuous variables, AGE, SYS and HRA do not have a cluster of values at 0.

Next, perform a bivariate analysis of categorical variables. Cross-classify each of the covariates with STA, produce chi-square statistics. An example of this analysis is shown below.

```
. tab sex sta, row col chi2

          | sta
      sex |         0          1 |     Total
----------+----------------------+----------
        0 |       100         24 |       124
          |     80.65      19.35 |    100.00
          |     62.50      60.00 |     62.00
----------+----------------------+----------
        1 |        60         16 |        76
          |     78.95      21.05 |    100.00
          |     37.50      40.00 |     38.00
----------+----------------------+----------
    Total |       160         40 |       200
          |     80.00      20.00 |    100.00
          |    100.00     100.00 |    100.00

          Pearson chi2(1) =    0.0849   Pr = 0.771
```

Perform a bivariate analysis of continuous variables using t-tests and Wilcoxon rank-sum tests. Note that none of the variables is particularly skewed so we expect agreement between the two tests. An example of this analysis is shown below.

```
. ttest hra, by(sta)

Two-sample t test with equal variances

------------------------------------------------------------------------------
   Group |     Obs        Mean    Std. Err.   Std. Dev.   [95% Conf. Interval]
---------+--------------------------------------------------------------------
       0 |     160        98.5    2.132852    26.97868    94.28763    102.7124
       1 |      40     100.625    4.188918    26.49304    92.15211    109.0979
---------+--------------------------------------------------------------------
combined |     200      98.925    1.897141    26.82962    95.18392    102.6661
---------+--------------------------------------------------------------------
    diff |              -2.125    4.752415               -11.49684    7.246844
------------------------------------------------------------------------------
Degrees of freedom: 198

                   Ho: mean(0) - mean(1) = diff = 0

   Ha: diff < 0               Ha: diff ~= 0               Ha: diff > 0
     t =  -0.4471               t =  -0.4471               t =  -0.4471
   P < t =   0.3276           P > |t| =   0.6553         P > t =   0.6724
```

```
. ranksum hra, by(sta)

Two-sample Wilcoxon rank-sum (Mann-Whitney) test

        sta |      obs    rank sum    expected
  ----------+---------------------------------
          0 |      160       15920       16080
          1 |       40        4180        4020
  ----------+---------------------------------
   combined |      200       20100       20100

  unadjusted variance   107200.00
  adjustment for ties      -61.91
                        ----------
  adjusted variance     107138.09

  Ho: hra(sta==0) = hra(sta==1)
              z =   -0.489
     Prob > |z| =    0.6250
```

Begin modeling by including all variables significant with p≤0.25. This model is shown below.

```
. logit sta ser crn inf cpr typ po2 bic cre loc2 age invsys

Iteration 0:   Log Likelihood =-100.08048
Iteration 1:   Log Likelihood =-72.218072
Iteration 2:   Log Likelihood =-70.217594
Iteration 3:   Log Likelihood =-70.079482
Iteration 4:   Log Likelihood =-70.077446
Iteration 5:   Log Likelihood =-70.077445

Logit Estimates                       Number of obs =      200
                                      chi2(11)      =    60.01
                                      Prob > chi2   =   0.0000
Log Likelihood = -70.077445           Pseudo R2     =   0.2998

------------------------------------------------------------------------------
    sta |     Coef.   Std. Err.       z      P>|z|    [95% Conf. Interval]
--------+---------------------------------------------------------------------
    ser | -.0978026   .5244226    -0.186    0.852    -1.125652    .9300469
    crn |  .4573635   .6938021     0.659    0.510    -.9024638    1.817191
    inf |   .176014   .4770485     0.369    0.712     -.758984    1.111012
    cpr |  .6625895   .8491518     0.780    0.435    -1.001717    2.326896
    typ |  1.889507   .862431      2.191    0.028      .199173    3.579841
    po2 |  .0173853   .6836773     0.025    0.980    -1.322598    1.357368
    bic | -.0191281   .7426827    -0.026    0.979     -1.47476    1.436503
    cre |  .5009578   .8804427     0.569    0.569    -1.224678    2.226594
   loc2 |  3.269501   .8926974     3.662    0.000     1.519846    5.019155
    age |  .0295018   .0128073     2.304    0.021        .0044    .0546037
 invsys |  142.0871   94.0869      1.510    0.131    -42.31979    326.4941
  _cons | -6.452996   1.415346    -4.559    0.000    -9.227022   -3.678969
------------------------------------------------------------------------------
```

Use the Wald statistics to delete variables one by one, least significant first. Repeat this process until all variables are significant. The model at the end of this repetitive process is shown below.

```
. logit sta typ loc2 age
Iteration 0:   Log Likelihood =-100.08048
Iteration 1:   Log Likelihood =-74.932982
Iteration 2:   Log Likelihood =-72.908149
Iteration 3:   Log Likelihood =-72.659837
Iteration 4:   Log Likelihood =-72.655321
Iteration 5:   Log Likelihood =-72.655319

Logit Estimates                                 Number of obs =      200
                                                chi2(3)       =    54.85
                                                Prob > chi2   = 0.0000
Log Likelihood = -72.655319                     Pseudo R2     = 0.2740

------------------------------------------------------------------------
    sta |     Coef.    Std. Err.       z     P>|z|    [95% Conf. Interval]
--------+---------------------------------------------------------------
    typ |  2.344696    .8068379    2.906    0.004    .7633226    3.926069
   loc2 |  3.456222    .8293496    4.167    0.000    1.830726    5.081717
    age |  .0337016     .012084    2.789    0.005    .0100174    .0573858
  _cons | -5.771535    1.146177   -5.035    0.000      -8.018   -3.525069
------------------------------------------------------------------------
```

At this point, allow each of the variables not in the model the opportunity to enter the model one by one. As each variable enters the model, evaluate its statistical significance using the Wald test. Also, assess whether the variable is a confounder of other variables in the model by calculating the additional variable's impact on the coefficients for the other variables in the model.

At the conclusion of this process, it appears that CAN and possibly SYS should be added to the model.

```
. logit sta typ loc2 age can sys
Iteration 0:   Log Likelihood =-100.08048
Iteration 1:   Log Likelihood = -70.91498
Iteration 2:   Log Likelihood =-68.096045
Iteration 3:   Log Likelihood =-67.812174
Iteration 4:   Log Likelihood =-67.805773
Iteration 5:   Log Likelihood =-67.805767

Logit Estimates                                 Number of obs =      200
                                                chi2(5)       =    64.55
                                                Prob > chi2   = 0.0000
Log Likelihood = -67.805767                     Pseudo R2     = 0.3225

------------------------------------------------------------------------
    sta |     Coef.    Std. Err.       z     P>|z|    [95% Conf. Interval]
--------+---------------------------------------------------------------
    typ |  2.963649    .9222993    3.213    0.001    1.155976    4.771323
   loc2 |  3.851796     .952827    4.042    0.000    1.984289    5.719302
    age |  .0365664    .0127428    2.870    0.004    .0115911    .0615418
    can |  2.138916    .8441223    2.534    0.011    .4844664    3.793365
    sys | -.0124485    .0067912   -1.833    0.067   -.0257591     .000862
  _cons | -5.111985    1.582666   -3.230    0.001   -8.213954   -2.010016
------------------------------------------------------------------------
```

SYS is not significant at the $p \leq 0.05$ level, it also does not appear to be a confounder of other relationships in the model. At this point SYS will be deleted from further analyses here. As always, there may be clinical reasons for including SYS or other covariates in the model.

```
. logit sta typ loc2 age can

Iteration 0:   Log Likelihood =-100.08048
Iteration 1:   Log Likelihood =-72.639697
Iteration 2:   Log Likelihood =-69.899311
Iteration 3:   Log Likelihood =-69.574481
Iteration 4:   Log Likelihood =-69.567324
Iteration 5:   Log Likelihood =-69.567319

Logit Estimates                                Number of obs =      200
                                               chi2(4)       =    61.03
                                               Prob > chi2   =   0.0000
Log Likelihood = -69.567319                    Pseudo R2     =   0.3049

-----------------------------------------------------------------------
    sta |     Coef.   Std. Err.       z     P>|z|    [95% Conf. Interval]
--------+--------------------------------------------------------------
    typ |  3.102182   .9185987     3.377    0.001    1.301762    4.902602
   loc2 |   3.70546   .8764735     4.228    0.000    1.987604    5.423317
    age |  .0371765   .0127701     2.911    0.004    .0121475    .0622054
    can |  2.097107   .8384739     2.501    0.012    .4537285    3.740486
  _cons | -6.869779   1.318819    -5.209    0.000   -9.454617   -4.284941
-----------------------------------------------------------------------
```

Use the method of fractional polynomials to evaluate the scale of the continuous covariate, AGE, in the model, while considering the other covariates

```
. fracpoly logit sta age typ loc2 can, compare
........,
-> gen age_1 = x^-.5 if _sample
-> gen age_2 = ln(x) if _sample
where: x = age/10

Iteration 0:   Log Likelihood =-100.08048
Iteration 1:   Log Likelihood =-72.269733
Iteration 2:   Log Likelihood =-69.538874
Iteration 3:   Log Likelihood =-69.208093
Iteration 4:   Log Likelihood =-69.200326
Iteration 5:   Log Likelihood = -69.20032

Logit Estimates                                Number of obs =      200
                                               chi2(5)       =    61.76
                                               Prob > chi2   =   0.0000
Log Likelihood =  -69.20032                    Pseudo R2     =   0.3086

-----------------------------------------------------------------------
    sta |     Coef.   Std. Err.       z     P>|z|    [95% Conf. Interval]
--------+--------------------------------------------------------------
  age_1 |  28.78382    18.7553     1.535    0.125    -7.97589    65.54353
  age_2 |   8.73054   4.736252     1.843    0.065    -.5523428   18.01342
    typ |  3.180197   .9351503     3.401    0.001    1.347336    5.013058
   loc2 |  3.820345   .8942928     4.272    0.000    2.067564    5.573127
    can |  2.223528   .8616683     2.580    0.010    .5346893    3.912367
  _cons | -32.22741   16.44774    -1.959    0.050   -64.46439    .0095601
-----------------------------------------------------------------------
Deviance:  138.401. Best powers of age among 44 models fit: -.5 0.

Fractional Polynomial Model Comparisons:

-----------------------------------------------------------------------
age             df      Deviance     Gain   P(term)  Powers
-----------------------------------------------------------------------
Not in model     0      149.448       --      --
Linear           1      139.135     0.000    0.001   1
m = 1            2      138.722     0.412    0.521   2
m = 2            4      138.401     0.734    0.851   -.5 0
-----------------------------------------------------------------------
```

It is appropriate to include AGE in its linear form.

Form all possible two-way interaction terms using the variables in the base model.

```
. gen typloc=typ*loc2

. gen typage=typ*age

. gen typcan=typ*can

. gen loc2age=loc2*age

. gen loc2can=loc2*can

. gen agecan=age*can
```

Build models containing the newly formed interaction terms and their component covariates. An example is shown below.

```
. logit sta typ can typcan

Iteration 0:   Log Likelihood =-100.08048
Iteration 1:   Log Likelihood =-92.269551
Iteration 2:   Log Likelihood =-91.069631
Iteration 3:   Log Likelihood =-90.963532
Iteration 4:   Log Likelihood =-90.960898
Iteration 5:   Log Likelihood =-90.960895

Logit Estimates                               Number of obs =      200
                                              chi2(3)       =    18.24
                                              Prob > chi2   =   0.0004
Log Likelihood = -90.960895                   Pseudo R2     =   0.0911


------------------------------------------------------------------------------
    sta  |      Coef.   Std. Err.       z     P>|z|     [95% Conf. Interval]
---------+--------------------------------------------------------------------
    typ  |   2.493437    1.03196     2.416    0.016     .4708326    4.516042
    can  |   .9718606   1.448604     0.671    0.502    -1.867351    3.811072
 typcan  |   .5510853   1.723283     0.320    0.749    -2.826487    3.928657
  _cons  |  -3.610918   1.013422    -3.563    0.000    -5.597189   -1.624647
------------------------------------------------------------------------------
```

None of the interaction terms is significant. Therefore, no interaction terms will be included in the final model. The preliminary final model (purposeful selection) is presented in the table below.

Table. Logistic Regression Analysis of Factors Associated with STA=1 from the ICU Data Set (Purposeful Selection of Covariates).

Variable	Coefficient	Standard Error	Odds Ratio (95% CI)
Emergency Admission (vs. Elective)	3.1022	0.9186	22.25 (3.68, 134.64)
Deep Stupor or Coma (vs. Neither)	3.7055	0.8765	40.67 (7.30, 226.63)
Cancer Part of Present Problem (vs. Not)	2.0971	0.8385	8.14 (1.57, 42.11)
Age (years)	0.0372	0.0128	1.45* (1.13, 1.86)

*Odds ratio for a 10-year change in age.

Stepwise Selection:

```
. sw logit sta  age sex ser can crn inf cpr sys hra pre typ fra po2 ph pco bic
> cre  loc2 ( race_1 race_2), forward pe(.2) pr (.9)
                      begin with empty model
p = 0.0000 <  0.2000   adding    loc2
p = 0.0092 <  0.2000   adding    typ
p = 0.0053 <  0.2000   adding    age
p = 0.0124 <  0.2000   adding    can
p = 0.0668 <  0.2000   adding    sys
p = 0.1335 <  0.2000   adding    pco
p = 0.0339 <  0.2000   adding    ph
p = 0.1996 <  0.2000   adding    pre

Logit Estimates                            Number of obs =      200
                                           chi2(8)       =    73.29
                                           Prob > chi2   =   0.0000
Log Likelihood = -63.435014                Pseudo R2     =   0.3662


------------------------------------------------------------------------------
    sta |      Coef.   Std. Err.       z     P>|z|      [95% Conf. Interval]
--------+---------------------------------------------------------------------
   loc2 |   4.62669   1.038581     4.455    0.000     2.591109    6.662272
    typ |   3.056427  .9464918     3.229    0.001     1.201337    4.911517
    age |   .0428465  .0141565     3.027    0.002     .0151003    .0705927
    can |   2.567817  .9002321     2.852    0.004     .803394     4.332239
    sys |  -.0147163  .007146     -2.059    0.039    -.0287223   -.0007103
    pco |  -2.229701  1.001073    -2.227    0.026    -4.191769   -.2676337
     ph |   1.936899  .9023165     2.147    0.032     .1683912    3.705407
    pre |   .7701213  .6004549     1.283    0.200    -.4067487    1.946991
  _cons |  -5.503533  1.624683    -3.387    0.001    -8.687852   -2.319213
------------------------------------------------------------------------------
```

The stepwise process using forward selection with backward elimination with pe set at 0.20 and pr set at 0.90 identifies a model similar to that developed using purposeful selection. In addition to the variables that the two methods have in common, the stepwise process also identifies four additional covariates: SYS, PCO, PH, PRE. A larger value of the p-value for entry can be used if one wishes to check for possible confounding of non-significant covariates. In this case when we use of pe = 0.25 only SEX is added to the model with $p = 0.223$.

At this point only the coefficient for PRE is not significant at the 0.05 level.

Since a rather flexible p-value to enter (0.20) was used, we will use likelihood ratio tests to compare the models selected at each step with the final model. Based on these tests, it seems justified to eliminate PRE from the model.

```
. logistic sta loc2 typ age can sys pco ph pre

Logit Estimates                              Number of obs =      200
                                             chi2(8)       =    73.29
                                             Prob > chi2   =   0.0000
Log Likelihood = -63.435014                  Pseudo R2     =   0.3662


---------------------------------------------------------------------
   sta | Odds Ratio   Std. Err.      z     P>|z|    [95% Conf. Interval]
-------+-------------------------------------------------------------
  loc2 |  102.1754    106.1174     4.455   0.000    13.34457   782.3261
   typ |  21.25149    20.11436     3.229   0.001    3.324559   135.8453
   age |  1.043778    .0147762     3.027   0.002    1.015215   1.073144
   can |  13.03733    11.73662     2.852   0.004    2.233107   76.11453
   sys |  .9853914    .0070416    -2.059   0.039    .9716863   .9992899
   pco |  .1075606    .107676     -2.227   0.026    .0151195   .765188
    ph |  6.937205    6.259554     2.147   0.032    1.183399   40.66658
   pre |  2.160028    1.296999     1.283   0.200    .6658115   7.007572
---------------------------------------------------------------------

. lrtest, saving (0)
```

```
. logistic sta loc2 typ age can sys pco ph

Logit Estimates                              Number of obs =      200
                                             chi2(7)       =    71.72
                                             Prob > chi2   =   0.0000
Log Likelihood = -64.220002                  Pseudo R2     =   0.3583


---------------------------------------------------------------------
   sta | Odds Ratio   Std. Err.      z     P>|z|    [95% Conf. Interval]
-------+-------------------------------------------------------------
  loc2 |  89.67835    90.94782     4.433   0.000    12.28674   654.5437
   typ |  20.64862    19.72269     3.170   0.002    3.175846   134.2526
   age |  1.043608    .0143047     3.114   0.002    1.015945   1.072025
   can |  11.04545    9.809187     2.705   0.007    1.937549   62.96717
   sys |  .9855667    .0069804    -2.053   0.040    .97198     .9993434
   pco |  .1050606    .1044717    -2.266   0.023    .0149627   .7376835
    ph |  6.387162    5.583454     2.121   0.034    1.151369   35.43246
---------------------------------------------------------------------

. lrtest
Logistic:   likelihood-ratio test               chi2(1)     =      1.57
                                                Prob > chi2 =    0.2102
```

Similar likelihood ratio tests suggest that the remaining variables are important to include in the model.

Use the method of fractional polynomials to re-evaluate the scale of the continuous covariates, AGE and SYS in the model, while considering the other covariates. These analyses suggest that it is appropriate to include both AGE and SYS in their original, linear forms.

Form all possible two-way interaction terms using the variables in the base model. Perform a second stepwise logistic regression to consider the interaction terms. In this model, include all of the main effects.

```
. sw logit sta (loc2 typ age can sys pco ph)  typloc typage typcan loc2age loc2
> can agecan loc2sys loc2pco loc2ph typsys typpco typph agesys agepco ageph can
> sys canpco canph syspco sysph pcoph, forward pr(.9) pe(.2)
(loc2can dropped because constant)
(loc2ph dropped because constant)
(loc2 dropped due to estimability)
(can dropped due to estimability)
(pco dropped due to estimability)
(typloc dropped due to estimability)
(typage dropped due to estimability)
(typcan dropped due to estimability)
(agecan dropped due to estimability)
(loc2pco dropped due to estimability)
(typsys dropped due to estimability)
(typph dropped due to estimability)
(cansys dropped due to estimability)
(canpco dropped due to estimability)
(canph dropped due to estimability)
(4 obs. dropped due to estimability)
                    begin with empty model
p = 0.0000 <  0.2000  adding     loc2age
p = 0.0016 <  0.2000  adding     typ age sys ph
p = 0.0217 <  0.2000  adding     syspco
p = 0.0258 <  0.2000  adding     agesys
p = 0.0935 <  0.2000  adding     loc2sys
p = 0.9920 >= 0.9000  removing   loc2age
p = 0.1658 <  0.2000  adding     pcoph

Logit Estimates                          Number of obs =      196
                                         chi2(8)       =    71.20
                                         Prob > chi2   =   0.0000
Log Likelihood = -63.578722              Pseudo R2     =   0.3589


------------------------------------------------------------------------------
     sta |     Coef.   Std. Err.       z     P>|z|    [95% Conf. Interval]
---------+--------------------------------------------------------------------
 loc2sys |   .0430971   .0139311     3.094   0.002     .0157927    .0704015
     typ |   2.544166     1.0059     2.529   0.011     .5726391    4.515694
     age |   .1541467   .0667353     2.310   0.021     .0233479    .2849455
     sys |   .0419489   .0329239     1.274   0.203    -.0225807    .1064786
      ph |   1.078878    1.00071     1.078   0.281    -.8824782    3.040234
  syspco |  -.0324822   .0148294    -2.190   0.028    -.0615473    -.003417
  agesys |  -.0009152   .0004999    -1.831   0.067    -.0018949    .0000646
   pcoph |    3.35891   2.423793     1.386   0.166    -1.391638    8.109458
    _cons |  -11.56639   4.584224    -2.523   0.012     -20.5513   -2.581474
------------------------------------------------------------------------------
```

Owing to a number of numerical problems with the interaction terms, the stepwise logistic regression is difficult to interpret. To evaluate the interaction terms, build models containing the main effects terms and each interaction term (one by one).

This method suggests that only the AGE*SYS interaction term is significant.

```
. logit sta loc2 typ age can sys pco ph agesys

Iteration 0:   Log Likelihood =-100.08048
Iteration 1:   Log Likelihood = -65.99123
Iteration 2:   Log Likelihood =-61.721663
Iteration 3:   Log Likelihood =-61.141088
Iteration 4:   Log Likelihood =-61.117207
Iteration 5:   Log Likelihood =-61.117142

Logit Estimates                                Number of obs =     200
                                               chi2(8)       =   77.93
                                               Prob > chi2   =  0.0000
Log Likelihood = -61.117142                    Pseudo R2     =  0.3893


---------------------------------------------------------------------------
    sta |     Coef.    Std. Err.        z     P>|z|    [95% Conf. Interval]
--------+------------------------------------------------------------------
   loc2 |   4.932408   1.187221      4.155    0.000     2.605497    7.259319
    typ |   3.168034   .9824882      3.225    0.001     1.242392    5.093675
    age |   .2031191   .0679219      2.990    0.003     .0699946    .3362437
    can |   2.604631   .9013441      2.890    0.004     .8380293    4.371233
    sys |   .0644739   .0321328      2.006    0.045     .0014947    .1274531
    pco |  -2.620771   1.027112     -2.552    0.011    -4.633874   -.6076681
     ph |   1.830244   .8673404      2.110    0.035     .130288      3.5302
 agesys |  -.001214    .0004875     -2.490    0.013    -.0021695   -.0002584
  _cons |  -15.94671   4.758488     -3.351    0.001    -25.27318   -6.620249
---------------------------------------------------------------------------
```

None of the other interaction terms is significant. Therefore, only the AGE*SYS interaction term will be included in the final model. The preliminary final model (stepwise selection) is presented in the table below.

Table. Logistic Regression Analysis of Factors Associated with STA=1 from the ICU Data Set (Stepwise Selection of Covariates).

Variable	Coefficient	Standard Error	Odds Ratio (95% CI)
Emergency Admission (vs. Elective)	3.1680	0.9186	22.25 (3.68, 134.64)
Deep Stupor or Coma (vs. Neither)	4.9324	0.8765	40.67 (7.30, 226.63)
Cancer Part of Present Problem (vs. Not)	2.6046	0.8385	8.14 (1.57, 42.11)
Age (years)	0.2031	0.0128	**
Systolic Blood Pressure (mm Hg)	0.0645	0.0321	**
Low PCO2 (vs. not)	-2.6208	1.0271	0.07 (0.01, 0.54)
Low PH (vs. not)	1.8302	0.8673	6.24 (1.14, 34.13)
AGE*SYS	-0.0012	0.0005	**

*Odds ratio for a 10-year change in age.

Best Subsets Selection:

Best subsets regression was performed in SAS using the method described on pages 129 and 130 of the text. First, logistic regression was performed using all potential covariates. The fitted values from this regression were saved and were used to create z_i and v_i. Next, linear regression was performed using z_i as the dependent variable and v_i as the weight variable. In this step, selection of the best five models based on Mallow's C_p was requested. The commands used are shown below.

```
proc logistic descending;
   model sta=age sex race_1 race_2 can crn cpr sys hra typ po2 ph pco loc2 ser inf
        pre fra bic cre;
   output out=alr1.new5 prob=pihat;
data alr1.new6;
set alr1.new5;
   z=log(pihat/(1-pihat))+((sta-pihat)/(pihat*(1-pihat)));
   v=pihat*(1-pihat);
proc reg;
   model z=age sex race_1 race_2 can crn cpr sys hra typ po2 ph pco loc2 ser inf
        pre fra bic cre/selection=cp best=5;
   weight v;
```

The above program was run, yielding the output shown below.

```
                    The LOGISTIC Procedure
                       Model Information

          Data Set                      ALR1.NEW2
          Response Variable             sta
          Number of Response Levels     2
          Number of Observations        200
          Link Function                 Logit
          Optimization Technique        Fisher's scoring


                       Response Profile

             Ordered                        Total
              Value         sta           Frequency

                1            1                40
                2            0               160


                  Model Convergence Status

          Convergence criterion (GCONV=1E-8) satisfied.


                    Model Fit Statistics

                                        Intercept
                         Intercept         and
          Criterion         Only        Covariates

           AIC            202.161         161.832
           SC             205.459         231.097
           -2 Log L       200.161         119.832


            Testing Global Null Hypothesis: BETA=0

         Test              Chi-Square     DF     Pr > ChiSq

         Likelihood Ratio   80.3287       20      <.0001
         Score              75.5087       20      <.0001
         Wald               28.7574       20      0.0926
```

Analysis of Maximum Likelihood Estimates

Parameter	DF	Estimate	Standard Error	Chi-Square	Pr > ChiSq
Intercept	1	-5.1754	2.3870	4.7009	0.0301
age	1	0.0567	0.0186	9.3079	0.0023
sex	1	-0.6532	0.5310	1.5128	0.2187
race_1	1	-1.6978	1.7596	0.9311	0.3346
race_2	1	-0.4237	1.2836	0.1090	0.7413
can	1	3.1703	1.0523	9.0771	0.0026
crn	1	-0.0787	0.8194	0.0092	0.9235
cpr	1	0.9574	1.0203	0.8805	0.3481
sys	1	-0.0152	0.00867	3.0946	0.0786
hra	1	-0.00167	0.00974	0.0293	0.8642
typ	1	3.0478	1.0901	7.8165	0.0052
po2	1	-0.1126	0.8723	0.0167	0.8973
ph	1	2.4376	1.2287	3.9360	0.0473
pco	1	-3.0807	1.2711	5.8743	0.0154
loc2	1	5.4450	1.3058	17.3867	<.0001
ser	1	-0.7861	0.6185	1.6153	0.2037
inf	1	-0.1979	0.5518	0.1286	0.7199
pre	1	1.1370	0.6792	2.8022	0.0941
fra	1	1.5566	1.0505	2.1956	0.1384
bic	1	-0.6778	0.9129	0.5512	0.4578
cre	1	0.1003	1.1183	0.0080	0.9285

The REG Procedure
Model: MODEL1
Dependent Variable: z

Weight: v

Analysis of Variance

Source	DF	Sum of Squares	Mean Square	F Value	Pr > F
Model	20	28.75820	1.43791	1.10	0.3497
Error	179	233.30773	1.30340		
Corrected Total	199	262.06593			

Root MSE	1.14166	R-Square	0.1097
Dependent Mean	-1.09155	Adj R-Sq	0.0103
Coeff Var	-104.59067		

```
                                Parameter Estimates

                        Parameter      Standard
          Variable    DF   Estimate       Error    t Value    Pr > |t|

          Intercept    1   -5.17546      2.72513     -1.90      0.0592
          age          1    0.05671      0.02122      2.67      0.0082
          sex          1   -0.65317      0.60628     -1.08      0.2828
          race_1       1   -1.69786      2.00885     -0.85      0.3991
          race_2       1   -0.42371      1.46543     -0.29      0.7728
          can          1    3.17039      1.20135      2.64      0.0090
          crn          1   -0.07865      0.93552     -0.08      0.9331
          cpr          1    0.95736      1.16479      0.82      0.4122
          sys          1   -0.01525      0.00989     -1.54      0.1251
          hra          1   -0.00167      0.01112     -0.15      0.8810
          typ          1    3.04786      1.24455      2.45      0.0153
          po2          1   -0.11256      0.99588     -0.11      0.9101
          ph           1    2.43764      1.40274      1.74      0.0840
          pco          1   -3.08072      1.45112     -2.12      0.0351
          loc2         1    5.44512      1.49084      3.65      0.0003
          ser          1   -0.78608      0.70611     -1.11      0.2671
          inf          1   -0.19788      0.63001     -0.31      0.7538
          pre          1    1.13696      0.77541      1.47      0.1443
          fra          1    1.55658      1.19929      1.30      0.1960
          bic          1   -0.67778      1.04222     -0.65      0.5163
          cre          1    0.10031      1.27675      0.08      0.9375

                             The REG Procedure
                              Model: MODEL1
                          Dependent Variable: z

                           C(p) Selection Method

                                Weight: v

Number in
  Model       C(p)    R-Square   Variables in Model

      4     -0.6556    0.0583    age can typ loc2
      5      0.0502    0.0647    age can sys typ loc2
      3      0.1369    0.0444    age typ loc2
      2      0.4902    0.0327    typ loc2
      5      0.5420    0.0623    age sex can typ loc2
```

Note that the model coefficients from the LOGISTIC and REG procedures are the same for each covariate. The best subsets selection indicates that the model containing AGE, CAN, TYP and LOC2 is the best model on the basis of C_p. These are the same main effects that were identified using the purposeful selection strategy.

The covariates AGE, CAN, TYP and LOC2 are identified as significant using all three modeling strategies. Stepwise selection also identified SYS, PCO and PH as important main effects. The small number of people in cells for PCO and PH leads to instability in the model reflected by large coefficients and standard errors, therefore these variables are excluded from the final model. The variable SYS and its signficant interaction with AGE, AGESYS, will be included in the final model. The final model for the ICU data is shown below.

Table. Logistic Regression Analysis of Factors Associated with STA=1 from the ICU Data Set, Final Model.

Variable	Coefficient	Standard Error	Odds Ratio (95% CI)
Emergency Admission (vs. Elective)	2.938	0.9185	18.89 (3.12, 114.28)
Deep Stupor or Coma (vs. Neither)	3.815	0.9404	45.39 (7.19, 286.70)
Cancer Part of Present Problem (vs. Not)	2.227	0.8410	9.28 (1.78, 48.22)
Age (years)	0.172	0.0663	**
Systolic Blood Pressure (mm Hg)	0.054	0.0312	**
AGE*SYS	-0.001	0.0005	**

Table. Odds Ratios for a 10-year increase in AGE at Various Levels of SYS.

	SYS=100	SYS=140	SYS=180
Odds Ratio for 10-year Increase in AGE	1.99	1.32	0.87
95% Confidence Interval	(1.29, 3.09)	(0.98, 1.77)	(0.52, 1.46)

4. Repeat Exercises 1 and 3 for the Low Birthweight Data and the Prostatic Cancer Data.

Low Birthweight Data

Consider the variable low birth weight (LOW) as the outcome variable and the remainder of the variables in the low birthweight data set as potential covariates. Use each of the variable selection methods discussed in this chapter to find a "best" model. Document thoroughly the rationale for each step in each process you follow. Compare and contrast the models resulting from the different approaches to variable selection. Note that in all cases the analysis should address not only identification of main effects but also appropriate scale for continuous covariates and potential interactions. Display the results of your final model in a table. Include in the table point and 95% CI estimates of all relevant odds ratios. Document the rationale for choosing the final model.

> *Purposeful Selection:*
>
> Begin with a univariate description of all covariates. An example of this analysis for a catagorical and a continuous variable is shown below.

```
. tab smoke

      smoke |      Freq.      Percent        Cum.
------------+-----------------------------------
          0 |        115        60.85       60.85
          1 |         74        39.15      100.00
------------+-----------------------------------
      Total |        189       100.00

. sum lwt, det

                             lwt
-------------------------------------------------------------
      Percentiles      Smallest
 1%           85             80
 5%           94             85
10%           98             85         Obs                189
25%          110             89         Sum of Wgt.        189

50%          121                        Mean          129.8148
                          Largest       Std. Dev.      30.57938
75%          140            229
90%          170            235         Variance      935.0985
95%          189            241         Skewness      1.390855
99%          241            250         Kurtosis      5.309181
```

> The univariate analysis does not reveal any variables for which there are illegal values. For each of the categorical variables, at least 1% of cases has a value greater than 0. All binary variables are coded as 0/1. Race (RACE), History of Premature Labor (PTL) and Number of Physician Visits During the First Trimester (FTV) are the only non-binary categorical variables. Indicator variables for RACE are created.

RACE	Label	RACE_2	RACE_3
1	White	0	0
2	Black	1	0
3	Other	0	1

The distribution of history of Premature Labor (PTL) is shown below. Since the number of women with a history of more than one premature labor is quite small, a dichotomous variable PTLD was created. PTLD takes on the value 1 if a woman has a history of one or more premature labors, and the value 0 if there is no history of premature labor. The cross-classification of the original variable PTL by the new variable PTLD is shown below.

```
. tab ptl

        ptl |      Freq.      Percent        Cum.
------------+-----------------------------------
          0 |        159        84.13       84.13
          1 |         24        12.70       96.83
          2 |          5         2.65       99.47
          3 |          1         0.53      100.00
------------+-----------------------------------
      Total |        189       100.00

. tab ptl ptld

            |        ptld
        ptl |         0          1 |     Total
------------+----------------------+----------
          0 |       159          0 |       159
          1 |         0         24 |        24
          2 |         0          5 |         5
          3 |         0          1 |         1
------------+----------------------+----------
      Total |       159         30 |       189
```

The distribution of number of physician visits during the first trimester (FTV) is shown below. Since the number of women with more than 2 physician visits during the first trimester is small, a three-level variable FTV3 was created. FTV3 takes on the value 0 if there were no physician visits during the first trimester, the value 1 if there was one physician visit during the first trimester and the value 2 if there were two or more visits during the first trimester. The cross-classification of the original variable FTV by the new variable FTV3 is shown below.

```
. tab ftv

        ftv |      Freq.      Percent        Cum.
------------+-----------------------------------
          0 |        100        52.91       52.91
          1 |         47        24.87       77.78
          2 |         30        15.87       93.65
          3 |          7         3.70       97.35
          4 |          4         2.12       99.47
          6 |          1         0.53      100.00
------------+-----------------------------------
      Total |        189       100.00
```

```
. tab ftv ftv3

        | ftv3
   ftv  |         0         1         2 |     Total
--------+-------------------------------+----------
     0  |       100         0         0 |       100
     1  |         0        47         0 |        47
     2  |         0         0        30 |        30
     3  |         0         0         7 |         7
     4  |         0         0         4 |         4
     6  |         0         0         1 |         1
--------+-------------------------------+----------
  Total |       100        47        42 |       189
```

The two continuous variables, AGE, and weight of the mother at the last menstrual period, LWT do not have a cluster of values at 0.

Next, perform a bivariate analysis of categorical variables. Cross-classify each of the covariates with LOW, produce chi-square statistics. An example of this analysis is shown below.

```
. tab smoke low, row col chi2

          | low
    smoke |         0         1 |     Total
----------+--------------------+----------
        0 |        86        29 |       115
          |     74.78     25.22 |    100.00
          |     66.15     49.15 |     60.85
----------+--------------------+----------
        1 |        44        30 |        74
          |     59.46     40.54 |    100.00
          |     33.85     50.85 |     39.15
----------+--------------------+----------
    Total |       130        59 |       189
          |     68.78     31.22 |    100.00
          |    100.00    100.00 |    100.00

          Pearson chi2(1) =    4.9237   Pr = 0.026
```

Perform a bivariate analysis of continuous variables using t-tests and Wilcoxon rank-sum tests. An example of this analysis is shown below.

```
. ttest lwt, by(low)

Two-sample t test with equal variances

------------------------------------------------------------------------------
   Group |     Obs        Mean    Std. Err.   Std. Dev.   [95% Conf. Interval]
---------+--------------------------------------------------------------------
       0 |     130       133.3     2.78238    31.72402     127.795     138.805
       1 |      59    122.1356    3.457723    26.55928    115.2142     129.057
---------+--------------------------------------------------------------------
combined |     189    129.8148    2.224323    30.57938     125.427    134.2027
---------+--------------------------------------------------------------------
    diff |             11.16441    4.743297                1.807158    20.52166
------------------------------------------------------------------------------
Degrees of freedom: 187

                  Ho: mean(0) - mean(1) = diff = 0

    Ha: diff < 0                 Ha: diff ~= 0                 Ha: diff > 0
      t =   2.3537                 t =   2.3537                 t =   2.3537
  P < t =   0.9902          P > |t| =   0.0196             P > t =   0.0098
```

```
. ranksum lwt, by (low)

Two-sample Wilcoxon rank-sum (Mann-Whitney) test

     low |      obs    rank sum    expected
---------+--------------------------------------
       0 |      130      13217.5       12350
       1 |       59       4737.5        5605
---------+--------------------------------------
combined |      189        17955       17955

unadjusted variance     121441.67
adjustment for ties       -181.97
                        ----------
adjusted variance       121259.70

Ho: lwt(low==0) = lwt(low==1)
           z =    2.491
    Prob > |z| =   0.0127
```

Begin modeling by including all variables significant with p≤0.25. This model is shown below.

```
. logit low smoke ptld ht ui race_2 race_3 ftv3 age lwt

Iteration 0:   Log Likelihood =  -117.336
Iteration 1:   Log Likelihood =-99.216558
Iteration 2:   Log Likelihood =-98.413605
Iteration 3:   Log Likelihood =-98.401662
Iteration 4:   Log Likelihood =-98.401658

Logit Estimates                           Number of obs  =      189
                                          chi2(9)        =    37.87
                                          Prob > chi2    =   0.0000
Log Likelihood = -98.401658               Pseudo R2      =   0.1614


--------------------------------------------------------------------------------
     low |     Coef.    Std. Err.       z     P>|z|     [95% Conf. Interval]
---------+----------------------------------------------------------------------
   smoke |   .8578743    .4130718     2.077    0.038     .0482684     1.66748
    ptld |   1.221774    .4629603     2.639    0.008     .3143886     2.12916
      ht |   1.85211     .7079552     2.616    0.009      .464543    3.239677
      ui |   .7141988    .4631075     1.542    0.123    -.1934752    1.621873
  race_2 |   1.217151    .5331813     2.283    0.022     .1721351    2.262167
  race_3 |   .8170808    .4542506     1.799    0.072    -.0732341    1.707396
    ftv3 |   .0397716    .2279801     0.174    0.862    -.4070613    .4866044
     age |  -.0389882    .038472     -1.013    0.311    -.1143919    .0364155
     lwt |  -.0149667    .0070431    -2.125    0.034    -.0287709   -.0011626
   _cons |   .6331242    1.2284       0.515    0.606    -1.774495    3.040744
--------------------------------------------------------------------------------
```

Use the Wald statistics to delete variables one by one that do not appear to be significant. Repeat this process until all variables are significant. The model at the end of this repetitive process is shown below.

```
. logit low smoke ptld ht   race_2 race_3   lwt

Iteration 0:   Log Likelihood =   -117.336
Iteration 1:   Log Likelihood =-100.82063
Iteration 2:   Log Likelihood =-100.24621
Iteration 3:   Log Likelihood =-100.24113
Iteration 4:   Log Likelihood =-100.24113

Logit Estimates                              Number of obs =      189
                                             chi2(6)       =    34.19
                                             Prob > chi2   = 0.0000
Log Likelihood = -100.24113                  Pseudo R2     = 0.1457


------------------------------------------------------------------------------
     low |     Coef.    Std. Err.       z     P>|z|     [95% Conf. Interval]
---------+--------------------------------------------------------------------
   smoke |   .8761063    .4007102     2.186    0.029     .0907287    1.661484
    ptld |   1.231437    .4462541     2.759    0.006     .3567947    2.106079
      ht |   1.767442     .70841      2.495    0.013     .3789844    3.155901
  race_2 |   1.263724    .5293288     2.387    0.017     .2262591     2.30119
  race_3 |   .8641763    .4350939     1.986    0.047     .0114079    1.716945
     lwt |  -.0167287    .0069499    -2.407    0.016    -.0303502   -.0031072
   _cons |   .0946195    .9570501     0.099    0.921    -1.781164    1.970403
------------------------------------------------------------------------------
```

At this point, allow each of the variables not in the model the opportunity to enter the model one by one. As each variable enters the model, evaluate its statistical significance using the Wald test. Also, assess whether the variable is a confounder of other variables in the model by calculating the additional variable's impact on the coefficients for the other variables in the model.

At the conclusion of this process, it appears while AGE is not statistically significant, it does impact the coefficients of the other variables in the model (in the 6-13% range). Since AGE is generally of clinical importance, it will be included in the model.

```
. logit low smoke ptld ht   race_2 race_3   lwt age

Iteration 0:   Log Likelihood =   -117.336
Iteration 1:   Log Likelihood =-100.32189
Iteration 2:   Log Likelihood =-99.585715
Iteration 3:   Log Likelihood =-99.575707
Iteration 4:   Log Likelihood =-99.575705

Logit Estimates                              Number of obs =      189
                                             chi2(7)       =    35.52
                                             Prob > chi2   = 0.0000
Log Likelihood = -99.575705                  Pseudo R2     = 0.1514


------------------------------------------------------------------------------
     low |     Coef.    Std. Err.       z     P>|z|     [95% Conf. Interval]
---------+--------------------------------------------------------------------
   smoke |   .8583324    .4047966     2.120    0.034     .0649456    1.651719
    ptld |    1.33397    .4575781     2.915    0.004     .4371332    2.230806
      ht |   1.740511    .7031128     2.475    0.013      .362435    3.118587
  race_2 |   1.168452    .5325845     2.194    0.028     .1246056    2.212299
  race_3 |   .8146204    .4427513     1.840    0.066    -.0531562    1.682397
     lwt |  -.0154359    .0070442    -2.191    0.028    -.0292422   -.0016296
     age |  -.0427843    .0375681    -1.139    0.255    -.1164165    .0308478
   _cons |   .9249097    1.202718     0.769    0.442    -1.432375    3.282194
------------------------------------------------------------------------------
```

Use the method of fractional polynomials to evaluate the scale of the continuous covariates, AGE and LWT in the model, while considering the other covariates.

```
. fracpoly logit low age lwt smoke ptld ht race_2 race_3, compare
........
-> gen age_1 = x^3 if _sample
-> gen age_2 = x^3*ln(x) if _sample
where: x = age/10

Iteration 0:  Log Likelihood =  -117.336
Iteration 1:  Log Likelihood =-100.16137
Iteration 2:  Log Likelihood =-98.982105
Iteration 3:  Log Likelihood =-98.768533
Iteration 4:  Log Likelihood =-98.765784
Iteration 5:  Log Likelihood =-98.765783

Logit Estimates                         Number of obs =      189
                                        chi2(8)       =    37.14
                                        Prob > chi2   = 0.0000
Log Likelihood = -98.765783             Pseudo R2     = 0.1583


------------------------------------------------------------------------
    low |    Coef.    Std. Err.      z     P>|z|     [95% Conf. Interval]
--------+---------------------------------------------------------------
  age_1 |  .2258286   .2541931    0.888    0.374    -.2723807    .7240379
  age_2 | -.1991024   .1996688   -0.997    0.319     -.590446    .1922412
    lwt | -.0156026   .0070695   -2.207    0.027    -.0294585   -.0017467
  smoke |  .8839374   .4080079    2.166    0.030     .0842565    1.683010
   ptld |  1.305261   .4621154    2.825    0.005     .3995312     2.21099
     ht |  1.725621   .7089752    2.434    0.015     .3360548    3.115186
 race_2 |  1.254191   .5365602    2.337    0.019     .2025525    2.30583
 race_3 |  .8004739   .4439035    1.803    0.071    -.0695609    1.670509
   _cons| -.590455    1.324396   -0.446    0.656    -3.186223    2.005313
------------------------------------------------------------------------
Deviance:  197.532. Best powers of age among 44 models fit: 3 3.

Fractional Polynomial Model Comparisons:
----------------------------------------------------------------
age             df     Deviance     Gain    P(term) Powers
----------------------------------------------------------------
Not in model     0      200.482      --        --
Linear           1      199.151    0.000     0.249  1
m = 1            2      198.584    0.568     0.451  3
m = 2            4      197.532    1.620     0.591  3 3
----------------------------------------------------------------
```

```
. fracpoly logit low lwt smoke ptld ht race_2 race_3 age, compare
........
-> gen lwt_1 = x^-2 if _sample
-> gen lwt_2 = x^3 if _sample
where: x = lwt/100

Iteration 0:   Log Likelihood =  -117.336
Iteration 1:   Log Likelihood =-100.05426
Iteration 2:   Log Likelihood =-99.320009
Iteration 3:   Log Likelihood =-99.308799
Iteration 4:   Log Likelihood =-99.308794

Logit Estimates                                   Number of obs =     189
                                                  chi2(8)       =   36.05
                                                  Prob > chi2   =  0.0000
Log Likelihood = -99.308794                       Pseudo R2     =  0.1536

------------------------------------------------------------------------------
    low |     Coef.   Std. Err.       z     P>|z|     [95% Conf. Interval]
--------+---------------------------------------------------------------------
  lwt_1 |   .8138864   1.025251     0.794   0.427    -1.195568    2.823341
  lwt_2 |  -.1357279   .1417917    -0.957   0.338    -.4136345    .1421787
  smoke |   .8436241    .411487     2.050   0.040     .0371243    1.650124
   ptld |   1.328836   .4583688     2.899   0.004     .4304499    2.227223
     ht |   1.766105   .7193553     2.455   0.014      .356195    3.176016
 race_2 |   1.176022   .5337929     2.203   0.028     .1298075    2.222237
 race_3 |   .8017803   .4478789     1.790   0.073    -.0760462    1.679607
    age |  -.0425851   .0375775    -1.133   0.257    -.1162357    .0310654
   _cons|  -1.276121   1.341856    -0.951   0.342    -3.906111    1.353868
------------------------------------------------------------------------------
Deviance:  198.618. Best powers of lwt among 44 models fit: -2 3.

Fractional Polynomial Model Comparisons:
----------------------------------------------------------------
lwt              df     Deviance     Gain    P(term) Powers
----------------------------------------------------------------
Not in model     0      204.552       --        --
Linear           1      199.151     0.000     0.020   1
m = 1            2      199.151     0.000     1.000   1
m = 2            4      198.618     0.534     0.766   -2 3
----------------------------------------------------------------
```

It is appropriate to include both AGE and LWT in their linear forms.

Form all possible two-way interaction terms using the variables in the base model.

```
. gen agelwt=age*lwt

. gen agesmk=age*smoke

. gen ageptld=age*ptld

. gen ageht=age*ht

. gen ager2=age*race_2

. gen ager3=age*race_3

. gen lwtsmk=lwt*smoke

. gen lwtptld=lwt*ptld

. gen lwtht=lwt*ht

. gen lwtr2=lwt*race_2

. gen lwtr3=lwt*race_3

. gen smkptld=smoke*ptld

. gen smkht=smoke*ht

. gen smkr2=smoke*race_2

. gen smkr3=smoke*race_3

. gen htr2=ht*race_2

. gen htr3=ht*race_3

. gen ptldht=ptld*ht

. gen ptldr2=ptld*race_2

. gen ptldr3=ptld*race_3
```

Build models containing the main effects and adding the newly formed interaction terms one by one. An example is shown below.

```
. logit low smoke ptld ht  race_2 race_3  lwt age  agelwt

Iteration 0:   Log Likelihood =  -117.336
Iteration 1:   Log Likelihood =-100.29026
Iteration 2:   Log Likelihood =-99.587552
Iteration 3:   Log Likelihood =-99.573157
Iteration 4:   Log Likelihood =-99.573144

Logit Estimates                              Number of obs =     189
                                             chi2(8)       =   35.53
                                             Prob > chi2   =  0.0000
Log Likelihood = -99.573144                  Pseudo R2     =  0.1514


------------------------------------------------------------------------------
    low |     Coef.    Std. Err.      z      P>|z|      [95% Conf. Interval]
--------+---------------------------------------------------------------------
  smoke |   .8571164   .4049501     2.117   0.034     .0634287    1.650804
   ptld |   1.336613   .4591216     2.911   0.004     .4367513    2.236475
     ht |    1.74053   .7031605     2.475   0.013      .362361    3.118699
 race_2 |   1.165093   .5346291     2.179   0.029     .1172394    2.212947
 race_3 |   .8114782   .4446248     1.825   0.068    -.0599704    1.682927
    lwt |  -.0175822   .0308105    -0.571   0.568    -.0779696    .0428053
    age |  -.0548182   .1721848    -0.318   0.750    -.3922942    .2826577
 agelwt |   .0000938   .0013096     0.072   0.943     -.002473    .0026607
   _cons |   1.200285    4.03127     0.298   0.766     -6.70086    9.101429
------------------------------------------------------------------------------
```

None of the interaction terms is significant. Therefore, no interaction terms will be included in the final model. The preliminary final model (purposeful selection) is presented in the table below.

Table. Logistic Regression Analysis of Factors Associated with LOW=1 from the Low Birthweight Data Set (Purposeful Selection of Covariates).

Variable	Coefficient	Standard Error	Odds Ratio (95% CI)
Smoking During Pregnancy (vs. No smoking)	0.8583	0.4048	2.36 (1.07, 5.22)
History of Premature Labor (vs. No History)	1.3340	0.4576	3.80 (1.55, 9.31)
History of Hypertension (vs. No History)	1.7405	0.7031	5.70 (1.44, 22.61)
Black Race (vs. White)	1.1685	0.5326	3.22 (1.13, 9.14)
Other Race (vs. White)	0.8146	0.4428	2.26 (0.95, 5.38)
Weight of Mother at Last Menstrual Period (pounds)	-0.0154	0.0070	0.86* (0.75, 0.98)
Age (years)	-0.0428	0.0376	0.65** (0.31, 1.36)

*Odds ratio for a 10-pound increase in weight.
**Odds ratio for a 10-year increase in age.

Stepwise Selection:

```
. sw logit low  age lwt (race_2 race_3) smoke ptld  ht ui ftvd, forward pe(.2)
> pr(.9)
                         begin with empty model
p = 0.0004 <  0.2000   adding    ptld
p = 0.0386 <  0.2000   adding    age
p = 0.0419 <  0.2000   adding    ht
p = 0.0250 <  0.2000   adding    lwt
p = 0.1277 <  0.2000   adding    ui
p = 0.1421 <  0.2000   adding    race_2 race_3
p = 0.0381 <  0.2000   adding    smoke

Logit Estimates                              Number of obs =      189
                                             chi2(8)       =    37.84
                                             Prob > chi2   = 0.0000
Log Likelihood = -98.416852                  Pseudo R2     = 0.1612

--------------------------------------------------------------------------
    low |     Coef.    Std. Err.       z     P>|z|     [95% Conf. Interval]
--------+-----------------------------------------------------------------
   ptld |   1.221751    .4630153    2.639    0.008     .3142574    2.129244
    age |  -.0377496    .0378109   -0.998    0.318    -.1118577    .0363584
     ht |   1.838687    .7032521    2.615    0.009     .4603384    3.217036
    lwt |  -.0149103    .0070405   -2.118    0.034    -.0287093   -.0011112
     ui |   .7111278    .4631199    1.536    0.125    -.1965706    1.618826
 race_2 |   1.212742    .5324883    2.277    0.023     .1690841      2.2564
 race_3 |   .8041194    .4484447    1.793    0.073     -.074816    1.683055
  smoke |   .8464023    .4080745    2.074    0.038     .0465909    1.646214
  _cons |   .6369097    1.230312    0.518    0.605    -1.774458    3.048277
--------------------------------------------------------------------------
```

The stepwise process using forward selection with backward elimination with pe set at 0.20 and pr set at 0.90 identifies a model similar to that developed using purposeful selection. In addition to the variables that the two methods have in common, the stepwise process also identifies one additional covariate: UI (history of uterine irritability).

Since a rather flexible p-value to enter (0.20) was used, we will use likelihood ratio tests to compare the models selected at each step with the final model. Based on these tests, it seems that there is some justification for the inclusion of UI in the model. A final decision about inclusion of this variable should be made taking into account its clinical significance.

```
. logistic low ptld age ht lwt ui race_2 race_3 smoke
Logit Estimates                              Number of obs =      189
                                             chi2(8)       =    37.84
                                             Prob > chi2   = 0.0000
Log Likelihood = -98.416852                  Pseudo R2     = 0.1612

--------------------------------------------------------------------------
    low | Odds Ratio   Std. Err.       z     P>|z|     [95% Conf. Interval]
--------+-----------------------------------------------------------------
   ptld |   3.393123    1.571068    2.639    0.008     1.369242    8.408509
    age |    .962954    .0364102   -0.998    0.318     .8941715    1.037027
     ht |   6.288278    4.422245    2.615    0.009      1.58461    24.95405
    lwt |   .9852003    .0069363   -2.118    0.034     .9716989    .9988894
     ui |   2.036286    .9430448    1.536    0.125     .8215433    5.047162
 race_2 |   3.362692    1.790594    2.277    0.023      1.18422     9.54865
 race_3 |   2.234728    1.002152    1.793    0.073     .9279142    5.381972
  smoke |   2.331245    .9513216    2.074    0.038     1.047693    5.187302
--------------------------------------------------------------------------
```

```
. lrtest, saving (0)

. logistic low ptld age ht lwt race_2 race_3 smoke

Logit Estimates                                    Number of obs =      189
                                                   chi2(7)       =    35.52
                                                   Prob > chi2   = 0.0000
Log Likelihood = -99.575705                        Pseudo R2     = 0.1514

------------------------------------------------------------------------------
     low | Odds Ratio   Std. Err.       z     P>|z|    [95% Conf. Interval]
---------+--------------------------------------------------------------------
    ptld | 3.796083    1.737005     2.915   0.004     1.548262    9.307368
     age |  .958118     .0359947   -1.139   0.255      .8901045   1.031329
      ht | 5.700254    4.007922     2.475   0.013     1.436824    22.61439
     lwt | .9846826     .0069363   -2.191   0.028      .9711812    .9983717
  race_2 | 3.217009    1.713329     2.194   0.028     1.132702    9.136693
  race_3 | 2.258318     .9998732    1.840   0.066      .9482319   5.378432
   smoke | 2.359223     .9550055    2.120   0.034     1.067101    5.215939
------------------------------------------------------------------------------

. lrtest
Logistic:  likelihood-ratio test                   chi2(1)       =     2.32
                                                   Prob > chi2   =   0.1279
```

Similar likelihood ratio tests suggest that the remaining variables are important to include in the model.

Use the method of fractional polynomials to re-evaluate the scale of the continuous covariates, AGE and LWT in the model, while considering the other covariates. These analyses suggest that it is appropriate to include both AGE and LWT in their original, linear forms.

Form all possible two-way interaction terms using the variables in the base model. Perform a second stepwise logistic regression to consider the interaction terms. In this model, include all of the main effects.

```
. sw logit low (smoke ptld ht race_2 race_3 lwt age ui)   agelwt agesmk ageptld
> ageht ager2 ager3 lwtsmk lwtptld lwtht lwtr2 lwtr3 smkptld smkht smkr2 smkr3
> ptldht ptldr2 ptldr3 htr2 htr3 smkui ptldui htui r2ui r3ui lwtui ageui, pe(.2
> ) pr(.9) forward
(htui dropped because constant)
(ht dropped due to estimability)
(ptldht dropped due to estimability)
(2 obs. dropped due to estimability)
                          begin with empty model
p = 0.0013 <   0.2000   adding    smkptld
p = 0.0139 <   0.2000   adding    agelwt
p = 0.0173 <   0.2000   adding    ageht
p = 0.0440 <   0.2000   adding    smkr2
p = 0.0340 <   0.2000   adding    lwtui

Logit Estimates                              Number of obs =      187
                                             chi2(5)       =    32.26
                                             Prob > chi2   =   0.0000
Log Likelihood = -98.853905                  Pseudo R2     =   0.1403


------------------------------------------------------------------------------
     low |     Coef.   Std. Err.      z     P>|z|     [95% Conf. Interval]
---------+--------------------------------------------------------------------
 smkptld |  1.675439   .5719037     2.930   0.003     .554528    2.796349
  agelwt | -.0006112   .0002075    -2.946   0.003    -.0010179   -.0002046
   ageht |  .0773726   .0321474     2.407   0.016     .0143649    .1403804
   smkr2 |  1.631182   .7366527     2.214   0.027     .1873696    3.074995
   lwtui |  .0078319   .0036951     2.120   0.034     .0005897    .0150741
   _cons |  .3991006   .5883884     0.678   0.498    -.7541194    1.552321
------------------------------------------------------------------------------
```

Owing to a number of numerical problems with the interaction terms, the stepwise logistic regression is difficult to interpret. To evaluate the interaction terms, build models containing the main effects terms and each interaction term (one by one).

None of the interaction terms is significant. The preliminary final model (stepwise selection) is presented in the table below.

Table. Logistic Regression Analysis of Factors Associated with LOW=1 from the Low Birthweight Data Set (Stepwise Selection of Covariates).

Variable	Coefficient	Standard Error	Odds Ratio (95% CI)
Smoking During Pregnancy (vs. No smoking)	0.8464	0.4081	2.33 (1.05, 5.19)
History of Premature Labor (vs. No History)	1.2218	0.4630	3.39 (1.40, 8.41)
History of Hypertension (vs. No History)	1.8387	0.7033	6.29 (1.58, 24.95)
Black Race (vs. White)	1.2127	0.5325	3.36 (1.18, 9.55)
Other Race (vs. White)	0.8041	0.4484	2.23 (0.93, 5.38)
Weight of Mother at Last Menstrual Period (pounds)	-0.0149	0.0070	0.86* (0.75, 0.99)
Age (years)	-0.0377	0.0378	0.69** (0.33, 1.44)
Presence of Uterine Irritability (vs. None)	0.7111	0.4631	2.04 (0.82, 5.05)

*Odds ratio for a 10-pound increase in weight
**Odds ratio for a 10-year increase in age

Best Subsets Selection:

Best subsets regression was performed in SAS using the method described on pages 129 and 130 of the text. First, logistic regression was performed using all potential covariates. The fitted values from this regression were saved and were used to create z_i and v_i. Next, linear regression was performed using z_i as the dependent variable and v_i as the weight variable. In this step, selection of the best five models based on Mallow's C_p was requested. The commands used are shown below.

```
proc logistic descending;
  model low=age lwt race_2 race_3 smoke ptld ht ui ftvd;
  output out=alr1.bw2 prob=pihat;
data alr1.bw3;
set alr1.bw2;
  z=log(pihat/(1-pihat))+((low-pihat)/(pihat*(1-pihat)));
  v=pihat*(1-pihat);
proc reg;
  model z=age lwt race_2 race_3 smoke ptld ht ui ftvd/selection=cp best=5;
  weight v;
run;
```

The above program was run, yielding the output shown below.

```
                    The LOGISTIC Procedure

                     Model Information

        Data Set                    ALR1.BW1
        Response Variable           low
        Number of Response Levels   2
        Number of Observations      189
        Link Function               Logit
        Optimization Technique      Fisher's scoring


                     Response Profile

          Ordered                        Total
          Value           low         Frequency

            1              1                59
            2              0               130


                Model Convergence Status

       Convergence criterion (GCONV=1E-8) satisfied.
```

```
                         Model Fit Statistics

                                              Intercept
                                 Intercept       and
             Criterion             Only       Covariates

             AIC                  236.672       216.729
             SC                   239.914       249.147
             -2 Log L             234.672       196.729


             Testing Global Null Hypothesis: BETA=0

        Test                Chi-Square      DF      Pr > ChiSq

        Likelihood Ratio       37.9429       9        <.0001
        Score                  35.5530       9        <.0001
        Wald                   28.6432       9        0.0007



            Analysis of Maximum Likelihood Estimates

                              Standard
    Parameter    DF   Estimate    Error   Chi-Square   Pr > ChiSq

    Intercept     1     0.6665    1.2380     0.2899       0.5903
    age           1    -0.0354    0.0385     0.8473       0.3573
    lwt           1    -0.0149    0.00705    4.4695       0.0345
    race_2        1     1.2022    0.5336     5.0759       0.0243
    race_3        1     0.7727    0.4597     2.8255       0.0928
    smoke         1     0.8148    0.4203     3.7578       0.0526
    ptld          1     1.2356    0.4656     7.0435       0.0080
    ht            1     1.8236    0.7053     6.6857       0.0097
    ui            1     0.7021    0.4646     2.2841       0.1307
    ftvd          1    -0.1215    0.3757     0.1045       0.7465




                      The REG Procedure
                       Model: MODEL1
                  Dependent Variable: z

                       Weight: v

                  Analysis of Variance

                              Sum of        Mean
    Source              DF    Squares      Square    F Value   Pr > F

    Model                9    28.64322    3.18258      3.16    0.0014
    Error              179   180.19751    1.00669
    Corrected Total    188   208.84073


            Root MSE              1.00334   R-Square     0.1372
            Dependent Mean       -0.66180   Adj R-Sq     0.0938
            Coeff Var          -151.60763
```

```
                            Parameter Estimates

                           Parameter      Standard
        Variable     DF    Estimate        Error     t Value    Pr > |t|

        Intercept    1      0.66650       1.24211       0.54     0.5922
        age          1     -0.03544       0.03863      -0.92     0.3602
        lwt          1     -0.01491       0.00707      -2.11     0.0365
        race_2       1      1.20224       0.53540       2.25     0.0260
        race_3       1      0.77270       0.46122       1.68     0.0956
        smoke        1      0.81484       0.42175       1.93     0.0549
        ptld         1      1.23555       0.46711       2.65     0.0089
        ht           1      1.82362       0.70763       2.58     0.0108
        ui           1      0.70213       0.46613       1.51     0.1338
        ftvd         1     -0.12146       0.37692      -0.32     0.7476

                           The REG Procedure
                            Model: MODEL1
                         Dependent Variable: z

                         C(p) Selection Method

                               Weight: v

   Number in
     Model         C(p)     R-Square     Variables in Model

         7        7.0917     0.1319      lwt race_2 race_3 smoke ptld ht ui
         6        7.6809     0.1194      lwt race_2 race_3 smoke ptld ht
         8        8.1038     0.1367      age lwt race_2 race_3 smoke ptld ht ui
         5        8.4101     0.1063      lwt race_2 ptld ht ui
         7        8.4355     0.1254      age lwt race_2 race_3 smoke ptld ht
```

Note that the model coefficients from the LOGISTIC and REG procedures are the same for each covariate. The best subsets selection indicates that the model containing LWT, RACE_2, RACE_3, SMOKE, PTLD, HT and UI is the best model on the basis of C_p. These are the same main effects that were identified using the stepwise selection strategy. Only the variable UI was not identified using the purposeful selection strategy.

The covariates LWT, RACE_2, RACE_3, SMOKE, PTLD, and HT are identified as significant using all three modeling strategies. AGE will be included in the final model as it is generally important for clinical reasons. The variable UI will not be included in the final model. The likelihood ratio test indicates that it is not statistically significant. Furthermore, its inclusion does not cause important changes in the coefficients for the other variables in the model. The final model for the low birthweight data is shown below.

Table. Logistic Regression Analysis of Factors Associated with LOW=1 from the Low Birthweight Data Set, Final Model.

Variable	Coefficient	Standard Error	Odds Ratio (95% CI)
Smoking During Pregnancy (vs. No smoking)	0.858	0.4048	2.36 (1.07, 5.22)
History of Premature Labor (vs. No History)	1.334	0.4576	3.80 (1.55, 9.31)
History of Hypertension (vs. No History)	1.741	0.7031	5.70 (1.44, 22.61)
Black Race (vs. White)	1.168	0.5326	3.22 (1.13, 9.14)
Other Race (vs. White)	0.815	0.4428	2.26 (0.95, 5.38)
Weight of Mother at Last Menstrual Period (pounds)	-0.015	0.0070	0.86* (0.75, 0.98)
Age (years)	-0.043	0.0376	0.65** (0.31, 1.36)

*Odds ratio for a 10-pound increase in weight.
**Odds ratio for a 10-year increase in age.

Prostatic Cancer Data

Consider the variable tumor penetration of the prostatic capsule (CAPSULE) as the outcome variable and the remainder of the variables in the prostatic cancer data set as potential covariates. Use each of the variable selection methods discussed in this chapter to find a "best" model. Document thoroughly the rationale for each step in each process you follow. Compare and contrast the models resulting from the different approaches to variable selection. Note that in all cases the analysis should address not only identification of main effects but also appropriate scale for continuous covariates and potential interactions. Display the results of your final model in a table. Include in the table point and 95% CI estimates of all relevant odds ratios. Document the rationale for choosing the final model.

Purposeful Selection:

Begin with a univariate description of all covariates. An example of this analysis for a categorical and a continuous variable is shown below.

```
. tab dpros

      dpros |      Freq.      Percent        Cum.
------------+-----------------------------------
        1 |         99        26.05       26.05
        2 |        132        34.74       60.79
        3 |         96        25.26       86.05
        4 |         53        13.95      100.00
------------+-----------------------------------
      Total |        380       100.00

. sum vol, det

                             vol
-------------------------------------------------------------
      Percentiles      Smallest
 1%            0              0
 5%            0              0
10%            0              0        Obs                 379
25%            0              0        Sum of Wgt.         379

50%         14.3                       Mean           15.85464
                       Largest         Std. Dev.      18.35381
75%         26.6           87.3
90%         38.3           87.6        Variance       336.8625
95%         48.7             96        Skewness       1.373583
99%         87.3           97.6        Kurtosis       5.485039
```

The univariate analysis does not reveal any variables for which there are illegal values. For each of the categorical variables, at least 1% of cases has a value greater than 0. RACE and DCAPS are originally coded as 1/2. These variables are recoded as 0/1. RACENEW takes on the value 0 when RACE =WHITE, and the value 1 when RACE=BLACK. DCAPSNEW takes on the value 0 when DCAPS=NO and the value 1 when DCAPS=YES. Results of the digital rectal exam, (DPROS) is the only non-binary categorical variable. Indicator variables for DPROS are created.

DPROS	Label	DPROS_2	DPROS_3	DPROS_4
1	No Nodule	0	0	0
2	Unilobar Nodule (Left)	1	0	0
3	Unilobar Nodule (Right)	0	1	0
4	Bilobar Nodule	0	0	1

The continuous variable, tumor volume obtained from ultrasound (VOL), has a cluster of values at 0. A binary variable VOL01 is created taking on the value 0 when VOL=0 and the value 1 when VOL=1. VOL01 will be included in any models containing VOL.

Next, perform a bivariate analysis of categorical variables. Cross-classify each of the covariates with CAPSULE, produce chi-square statistics. An example of this analysis is shown below.

```
. tab dpros capsule, row col chi2

          | capsule
   dpros  |         0          1 |     Total
----------+----------------------+----------
       1  |        80         19 |        99
          |     80.81      19.19 |    100.00
          |     35.24      12.42 |     26.05
----------+----------------------+----------
       2  |        84         48 |       132
          |     63.64      36.36 |    100.00
          |     37.00      31.37 |     34.74
----------+----------------------+----------
       3  |        45         51 |        96
          |     46.88      53.12 |    100.00
          |     19.82      33.33 |     25.26
----------+----------------------+----------
       4  |        18         35 |        53
          |     33.96      66.04 |    100.00
          |      7.93      22.88 |     13.95
----------+----------------------+----------
   Total  |       227        153 |       380
          |     59.74      40.26 |    100.00
          |    100.00     100.00 |    100.00

        Pearson chi2(3) =   40.3516    Pr = 0.000
```

Perform a bivariate analysis of continuous variables using t-tests and Wilcoxon rank-sum tests. An example of this analysis is shown below.

```
. ttest psa, by(capsule)

Two-sample t test with equal variances

------------------------------------------------------------------------------
   Group |     Obs        Mean    Std. Err.    Std. Dev.    [95% Conf. Interval]
---------+--------------------------------------------------------------------
       0 |     227    9.974405    .6887922      10.3777     8.617129    11.33168
       1 |     153    23.47118    2.18052      26.97154     19.16314    27.77922
---------+--------------------------------------------------------------------
combined |     380    15.40863    1.025854     19.99757     13.39155    17.42571
---------+--------------------------------------------------------------------
    diff |             -13.49677   1.976131                 -17.38236   -9.611184
------------------------------------------------------------------------------
Degrees of freedom: 378

                   Ho: mean(0) - mean(1) = diff = 0

     Ha: diff < 0                 Ha: diff ~= 0                Ha: diff > 0
       t =  -6.8299                t =  -6.8299                 t =  -6.8299
     P < t =   0.0000          P > |t| =   0.0000            P > t =   1.0000

. ranksum psa, by(capsule)

Two-sample Wilcoxon rank-sum (Mann-Whitney) test

 capsule |     obs    rank sum    expected
---------+-------------------------------
       0 |     227     36234.5     43243.5
       1 |     153     36155.5     29146.5
---------+-------------------------------
combined |     380       72390       72390

unadjusted variance    1102709.25
adjustment for ties        -65.84
                       ----------
adjusted variance      1102643.41

Ho: psa(capsule==0) = psa(capsule==1)
            z =   -6.675
    Prob > |z| =    0.0000
```

Begin modeling by including all variables. This model is shown below.

```
. logit capsule age racenew  dpros_2 dpros_3 dpros_4 dcapsnew vol vol01 logpsa
> gleason

Iteration 0:   Log Likelihood =-253.29367
Iteration 1:   Log Likelihood =-191.54838
Iteration 2:   Log Likelihood =-186.21478
Iteration 3:   Log Likelihood =-186.01109
Iteration 4:   Log Likelihood =-186.01062

Logit Estimates                              Number of obs =      376
                                             chi2(10)      =   134.57
                                             Prob > chi2   =   0.0000
Log Likelihood = -186.01062                  Pseudo R2     =   0.2656


------------------------------------------------------------------------------
 capsule |      Coef.    Std. Err.        z      P>|z|    [95% Conf. Interval]
---------+--------------------------------------------------------------------
     age | -.0118693    .0199956     -0.594    0.553    -.0510599     .0273213
 racenew |  -.644618      .46114     -1.398    0.162    -1.548436     .2591999
 dpros_2 |  .7596506     .359236      2.115    0.034     .0555609      1.46374
 dpros_3 |  1.541106    .3797444      4.058    0.000     .7968212     2.285392
 dpros_4 |  1.469021    .4620444      3.179    0.001     .5634306     2.374611
dcapsnew |  .4674877    .4627307      1.010    0.312    -.4394477     1.374423
     vol |  -.003463    .0123231     -0.281    0.779    -.0276158     .0206899
   vol01 | -.3862792    .4250157     -0.909    0.363    -1.219295     .4467363
  logpsa |  .5321723    .1608576      3.308    0.001     .2168972     .8474475
 gleason |  .9132594    .1689884      5.404    0.000     .5820481     1.244471
   _cons | -7.347851    1.630974     -4.505    0.000     -10.5445    -4.151201
------------------------------------------------------------------------------
```

Use the Wald statistics to delete variables one by one that do not appear to be significant. Repeat this process until all variables are significant. The model at the end of this repetitive process is shown below.

```
. logit capsule  dpros_2 dpros_3 dpros_4 logpsa gleason vol01

Iteration 0:   Log Likelihood =-256.14442
Iteration 1:   Log Likelihood =-193.61543
Iteration 2:   Log Likelihood =-188.28116
Iteration 3:   Log Likelihood =-188.08771
Iteration 4:   Log Likelihood =-188.08731

Logit Estimates                              Number of obs =      380
                                             chi2(6)       =   136.11
                                             Prob > chi2   =   0.0000
Log Likelihood = -188.08731                  Pseudo R2     =   0.2657


------------------------------------------------------------------------------
 capsule |      Coef.    Std. Err.        z      P>|z|    [95% Conf. Interval]
---------+--------------------------------------------------------------------
 dpros_2 |  .7833795    .3572313      2.193    0.028      .083219      1.48354
 dpros_3 |  1.610232    .3743363      4.302    0.000     .8765462     2.343918
 dpros_4 |  1.507925    .4495056      3.355    0.001     .6269102      2.38894
  logpsa |  .5153503    .1548282      3.329    0.001     .2118925      .818808
 gleason |  .9432278    .1647918      5.724    0.000     .6202417     1.266214
   vol01 | -.5733064    .2570519     -2.230    0.026    -1.077119    -.0694939
   _cons | -8.287855    1.055269     -7.854    0.000    -10.35614    -6.219566
------------------------------------------------------------------------------
```

At this point, allow each of the variables not in the model the opportunity to enter the model one by one. As each variable enters the model, evaluate its statistical significance using the Wald test. Also, assess whether the variable is a confounder of other variables in the model by calculating the additional variable's impact on the coefficients for the other variables in the model.

At the conclusion of this process, it does not appear that any of the originally excluded variables are significant. Furthermore none of the excluded variables appears to be a confounder of the relationships between the included variables and CAPSULE.

Use the method of fractional polynomials to re-evaluate the scale of the continuous covariates, PSA and GLEASON in the model, while considering the other covariates.

```
. fracpoly logit capsule psa  gleason dpros_2 dpros_3 dpros_4 vol01, compare
........
-> gen psa_1 = x^-1 if _sample
-> gen psa_2 = x^3 if _sample
where: x = psa/100

Iteration 0:  Log Likelihood =-256.14442
Iteration 1:  Log Likelihood =-196.02043
Iteration 2:  Log Likelihood =-188.15178
Iteration 3:  Log Likelihood =-186.72656
Iteration 4:  Log Likelihood =-186.57283
Iteration 5:  Log Likelihood =-186.57004
Iteration 6:  Log Likelihood =-186.57004

Logit Estimates                              Number of obs =     380
                                             chi2(7)       =  139.15
                                             Prob > chi2   =  0.0000
Log Likelihood = -186.57004                  Pseudo R2     =  0.2716


------------------------------------------------------------------------------
  capsule |    Coef.   Std. Err.      z     P>|z|    [95% Conf. Interval]
----------+-------------------------------------------------------------------
    psa_1 | -.0223488  .0103435   -2.161   0.031    -.0426217   -.0020759
    psa_2 |  4.002976  2.283261    1.753   0.080    -.4721331    8.478086
  gleason |  .9515862  .1623576    5.861   0.000     .6333711    1.269801
  dpros_2 |  .8147952  .3589922    2.270   0.023     .1111835    1.518407
  dpros_3 |  1.61659   .3750714    4.310   0.000     .8814635    2.351716
  dpros_4 |  1.502195  .4528104    3.317   0.001     .6147029    2.389687
    vol01 | -.5800633  .2576579   -2.251   0.024    -1.085064   -.0750631
    _cons | -6.931611  1.139059   -6.085   0.000    -9.164125   -4.699097
------------------------------------------------------------------------------
Deviance:  373.140. Best powers of psa among 44 models fit: -1 3.

Fractional Polynomial Model Comparisons:
----------------------------------------------------------------
psa             df      Deviance     Gain   P(term) Powers
----------------------------------------------------------------
Not in model     0      388.057       --       --
Linear           1      377.557     0.000    0.001   1
m = 1            2      376.175     1.382    0.240   0
m = 2            4      373.140     4.416    0.219   -1 3
----------------------------------------------------------------
```

```
. fracpoly logit capsule gleason  dpros_2 dpros_3 dpros_4 psa vol01, compare
........
-> gen gleaso_1 = x^-2 if _sample
-> gen gleaso_2 = x^3 if _sample
where: x = (gleason+1)

Iteration 0:  Log Likelihood =-256.14442
Iteration 1:  Log Likelihood =-194.62155
Iteration 2:  Log Likelihood =-190.10854
Iteration 3:  Log Likelihood =-189.85325
Iteration 4:  Log Likelihood = -189.7934
Iteration 5:  Log Likelihood =-189.59978
Iteration 6:  Log Likelihood =-188.53624
Iteration 7:  Log Likelihood =-188.51333
Iteration 8:  Log Likelihood =-188.51333

Logit Estimates                              Number of obs =      380
                                             chi2(7)       = 135.26
                                             Prob > chi2   = 0.0000
Log Likelihood = -188.51333                  Pseudo R2     = 0.2640


-------------------------------------------------------------------------
 capsule |     Coef.   Std. Err.       z     P>|z|     [95% Conf. Interval]
---------+---------------------------------------------------------------
gleaso_1 | -134.1466   82.15591    -1.633    0.103    -295.1693    26.87598
gleaso_2 |  .0019167   .0022798     0.841    0.400    -.0025516    .0063851
 dpros_2 |  .7587123   .3575745     2.122    0.034     .0578792    1.459545
 dpros_3 |  1.617376   .3760086     4.301    0.000     .8804122    2.354339
 dpros_4 |   1.46011   .4540539     3.216    0.001     .5701805    2.350039
     psa |  .0283401   .0094611     2.995    0.003     .0097966    .0468835
   vol01 | -.5437459    .256183    -2.122    0.034    -1.045855   -.0416364
   _cons |  .2277661   2.480462     0.092    0.927    -4.633851    5.089383
-------------------------------------------------------------------------

Note: 2 failures and 0 successes completely determined.
Deviance:  377.027. Best powers of gleason among 44 models fit: -2 3.

Fractional Polynomial Model Comparisons:
----------------------------------------------------------------
gleason           df    Deviance    Gain    P(term) Powers
----------------------------------------------------------------
Not in model       0     424.173     --       --
Linear             1     377.557    0.000    0.000   1
m = 1              2     377.160    0.397    0.529   -.5
m = 2              4     377.027    0.530    0.936   -2 3
----------------------------------------------------------------
```

It is appropriate to include both PSA and GLEASON in their linear forms.

Form all possible two-way interaction terms using the variables in the base model.

```
. gen d2psa= dpros_2*psa

. gen d3psa= dpros_3*psa

. gen d4psa= dpros_4*psa

. gen d2gle= dpros_2*gleason

. gen d3gle= dpros_3*gleason

. gen d4gle=dpros_4*gleason

. gen psagle=psa*gleason

. gen d2vol= dpros_2*vol01

. gen d3vol= dpros_3*vol01

. gen d4vol= dpros_4*vol01

. gen glevol=gleason*vol01

. gen psavol=psa*vol01
```

Build models containing the main effects and adding the newly formed interaction terms one by one. An example is shown below.

```
. logit capsule  dpros_2 dpros_3 dpros_4 gleason vol01 psa  d2vol d3vol d4vol

Iteration 0:   Log Likelihood =-256.14442
Iteration 1:   Log Likelihood =-190.51219
Iteration 2:   Log Likelihood =-183.87786
Iteration 3:   Log Likelihood =-183.54079
Iteration 4:   Log Likelihood =-183.53939
Iteration 5:   Log Likelihood =-183.53939

Logit Estimates                             Number of obs =     380
                                            chi2(9)       = 145.21
                                            Prob > chi2   = 0.0000
Log Likelihood = -183.53939                 Pseudo R2     = 0.2835

------------------------------------------------------------------------------
 capsule |     Coef.   Std. Err.      z      P>|z|    [95% Conf. Interval]
---------+--------------------------------------------------------------------
 dpros_2 |   1.662988   .5481003     3.034   0.002     .5887308    2.737245
 dpros_3 |   1.701331   .5944517     2.862   0.004     .5362268    2.866435
 dpros_4 |   2.626849    .763743     3.439   0.001      1.12994    4.123758
 gleason |   1.019609   .1670107     6.105   0.000     .6922741    1.346944
   vol01 |   .4416244   .5943503     0.743   0.457    -.7232808     1.60653
     psa |   .0281999   .0097807     2.883   0.004     .0090302    .0473697
    d2vol |  -1.743406   .7367117    -2.366   0.018    -3.187334   -.2994773
    d3vol |  -.2475666   .7579439    -0.327   0.744    -1.733109    1.237976
    d4vol |  -2.039905   .9734367    -2.096   0.036    -3.947805   -.1320036
    _cons |   -8.57321   1.172795    -7.310   0.000    -10.87185   -6.274575
------------------------------------------------------------------------------
```

There is a significant interaction between DPROS and VOL. Therefore, the interaction terms representing this interaction (D2VOL, D3VOL and D4VOL) will be included in the model. The preliminary final model (purposeful selection) is presented in the table below.

Table. Logistic Regression Analysis of Factors Associated with CAPSULE=1 from the Prostatic Cancer Data Set (Purposeful Selection of Covariates).

Variable	Coefficient	Standard Error	Odds Ratio (95% CI)
Unilobar Nodule Left (vs. No Nodule)	1.6630	0.5481	*
Unilobar Nodule Right (vs. No Nodule)	1.7013	0.5945	*
Bilobar Nodule (vs. No Nodule)	2.6268	0.7637	*
Tumor Volume >1cm^3 (vs. Tumor Vol=0)	0.4416	0.5944	*
Prostatic Specific Antigen Value (mg/ml)	0.0282	0.0098	1.36** (1.09, 1.61)
Total Gleason Score	1.020	0.1670	2.78*** (2.00, 3.85)
Unilobar Nodule Left*Volume	-1.7434	0.7367	*
Unilobar Nodule Right*Volume	-0.2476	0.7579	*
Bilobar Nodule*Volume	-2.0399	0.9734	*

*Odds ratios not computed for interacting term. See separate table.
**Odds ratio for a 10-unit increase in PSA value.
***Odds ratio for a 1-unit increase in GLEASON score.

Table. Odds Ratios for Results of Digital Rectal Exam for Subjects with Tumor Volume Obtained from Ultrasound = 0 and Subjects with Tumor Volume Obtained from Ultrasound >0.

	Odds Ratio (95% Confidence Interval)	
	Volume=1	Volume=0
Unilobular Nodule Left (vs. No Nodule)	0.92 (0.38, 2.41)	5.28 (1.80, 15.44)
Unilobular Nodule Right (vs. No Nodule)	4.28 (1.69, 10.86)	5.48 (1.71, 17.57)
Bilobar Nodule (vs. No Nodule)	1.80 (0.68, 4.76)	13.83 (3.1, 61.79)

It would also be appropriate to report the odds ratio for VOL01=1 vs. VOL01=0 at each level of DPROS. The confidence intervals suggest that there is considerable uncertainty in these estimates.

Stepwise Selection:

```
. sw logit capsule  age racenew ( dpros_2 dpros_3 dpros_4)   dcapsnew psa (vol01
>  vol) gleason, forward pe(.2) pr(.9)
                     begin with empty model
p = 0.0000 <  0.2000  adding    gleason
p = 0.0001 <  0.2000  adding    dpros_2 dpros_3 dpros_4
p = 0.0036 <  0.2000  adding    psa
p = 0.1070 <  0.2000  adding    vol01 vol

Logit Estimates                                Number of obs =     376
                                               chi2(7)       = 129.98
                                               Prob > chi2   = 0.0000
Log Likelihood =  -188.3046                    Pseudo R2     = 0.2566


------------------------------------------------------------------------------
 capsule |     Coef.   Std. Err.       z     P>|z|     [95% Conf. Interval]
---------+--------------------------------------------------------------------
 gleason |   .9850798   .1628425    6.049   0.000     .6659143    1.304245
 dpros_2 |   .7587116   .3580122    2.119   0.034     .0570206    1.460403
 dpros_3 |   1.592116   .3768841    4.224   0.000     .8534368    2.330795
 dpros_4 |   1.434651   .4569764    3.139   0.002     .5389938    2.330308
     psa |   .0277527   .0095693    2.900   0.004     .0089972    .0465082
   vol01 |   -.441545    .418238   -1.056   0.291    -1.261276    .3781864
     vol |  -.0034909   .0121582   -0.287   0.774    -.0273206    .0203387
    _cons |  -7.793856   1.072307   -7.268   0.000    -9.895539   -5.692173
------------------------------------------------------------------------------
```

The stepwise process using forward selection with backward elimination with pe set at 0.20 and pr set at 0.90 identifies a model similar to that developed using purposeful selection. In addition to the variables that the two methods have in common, the stepwise process also identifies one additional covariate: VOL (linear value of tumor volume). The stepwise regression is set up to include or exclude VOL and VOL01 together (to ensure that VOL is not included without VOL01).

Since a rather flexible p-value to enter (0.20) was used, we will use likelihood ratio tests to compare the models selected at each step with the final model. It is especially interesting to compare the models with and without VOL. Based on these tests, it seems that there is justification for the inclusion of VOL01 in the model; however VOL will not be included.

```
. logistic capsule gleason dpros_2 dpros_3 dpros_4 psa vol01 vol

Logit Estimates                             Number of obs =      379
                                            chi2(7)       = 133.86
                                            Prob > chi2   = 0.0000
Log Likelihood = -188.69746                 Pseudo R2     = 0.2618

-----------------------------------------------------------------------
capsule | Odds Ratio   Std. Err.       z     P>|z|    [95% Conf. Interval]
---------+-------------------------------------------------------------
gleason |   2.690695    .4385409     6.073   0.000    1.954934    3.703368
dpros_2 |   2.123943    .7614765     2.101   0.036    1.051889    4.288602
dpros_3 |   4.967671   1.873736      4.250   0.000    2.371873   10.40433
dpros_4 |   4.235373   1.935924      3.158   0.002    1.729102   10.37439
    psa |   1.028762    .0098298     2.968   0.003    1.009675    1.04821
  vol01 |    .6395395    .2674139    -1.069   0.285     .281806    1.451392
    vol |    .9964589    .0121342    -0.291   0.771     .9729579    1.020528
-----------------------------------------------------------------------

. lrtest, saving(0)

. logistic capsule gleason  dpros_2 dpros_3 dpros_4 psa vol01

Logit Estimates                             Number of obs =      380
                                            chi2(6)       = 134.73
                                            Prob > chi2   = 0.0000
Log Likelihood = -188.77826                 Pseudo R2     = 0.2630

-----------------------------------------------------------------------
capsule | Odds Ratio   Std. Err.       z     P>|z|    [95% Conf. Interval]
---------+-------------------------------------------------------------
gleason |   2.701307    .4396663     6.106   0.000    1.963504    3.716345
dpros_2 |   2.137791    .7661747     2.120   0.034    1.059006    4.315509
dpros_3 |   5.033994   1.891268      4.302   0.000    2.410554   10.51256
dpros_4 |   4.304612   1.958213      3.209   0.001    1.764874   10.49916
    psa |   1.028446    .009733      2.964   0.003    1.009546    1.047701
  vol01 |    .5798292    .1485122    -2.128   0.033     .3509789     .9578976
-----------------------------------------------------------------------

. lrtest
Warning:  observations differ:  379 vs. 380
Logistic:  likelihood-ratio test        chi2(1)    =         0.16
                                        Prob > chi2 =       0.6877
```

Similar likelihood ratio tests suggest that the remaining variables are important to include in the model.

Use the method of fractional polynomials to re-evaluate the scale of the continuous covariates, PSA and GLEASON in the model, while considering the other covariates. These analyses suggest that it is appropriate to include both PSA and GLEASON in their original, linear forms.

Form all possible two-way interaction terms using the variables in the base model. Perform a second stepwise logistic regression to consider the interaction terms. In this model, include all of the main effects.

```
. sw logit capsule (dpros_2 dpros_3 dpros_4 gleason psa vol01)  (d2psa d3psa d4
> psa) (d2gle d3gle d4gle) psagle (d2vol d3vol d4vol) glevol psavol, forward pe
> (.2) pr(.9)
                        begin with empty model
p = 0.0000 <  0.2000  adding   dpros_2 dpros_3 dpros_4 gleason psa vol01
p = 0.0179 <  0.2000  adding   d2vol d3vol d4vol

Logit Estimates                              Number of obs =      380
                                             chi2(9)       = 145.21
                                             Prob > chi2   = 0.0000
Log Likelihood = -183.53939                  Pseudo R2     = 0.2835


------------------------------------------------------------------------------
capsule |     Coef.    Std. Err.      z     P>|z|    [95% Conf. Interval]
--------+---------------------------------------------------------------------
dpros_2 |   1.662988    .5481003    3.034   0.002     .5887308    2.737245
dpros_3 |   1.701331    .5944517    2.862   0.004     .5362268    2.866435
dpros_4 |   2.626849    .763743     3.439   0.001     1.12994     4.123758
gleason |   1.019609    .1670107    6.105   0.000     .6922741    1.346944
    psa |   .0281999    .0097807    2.883   0.004     .0090302    .0473697
  vol01 |   .4416244    .5943503    0.743   0.457    -.7232808    1.60653
  d2vol |  -1.743406    .7367117   -2.366   0.018    -3.187334   -.2994773
  d3vol |  -.2475666    .7579439   -0.327   0.744    -1.733109    1.237976
  d4vol |  -2.039905    .9734367   -2.096   0.036    -3.947805   -.1320036
   _cons |  -8.57321    1.172795   -7.310   0.000    -10.87185   -6.274575
------------------------------------------------------------------------------
```

The stepwise logistic regression identifies the same interaction terms as the purposefully selected model. There is a significant interaction between DPROS and VOL01. Therefore, the interaction terms representing this interaction (D2VOL, D3VOL and D4VOL) will be included in the model. The preliminary final model from stepwise selection is the same as the model from purposeful selection, which has already been presented.

Best Subsets Selection:

Best subsets regression was performed in SAS using the method described on pages 129 and 130 of the text. First, logistic regression was performed using all potential covariates. The fitted values from this regression were saved and were used to create z_i and v_i. Next, linear regression was performed using z_i as the dependent variable and v_i as the weight variable. In this step, selection of the best five models based on Mallow's C_p was requested. The commands used are shown below.

```
proc logistic descending;
   model capsule=age racenew dcapsnew dpros_2 dpros_3 dpros_4 vol vol01 gleason;
   output out=alr1.pnew2 prob=pihat;
data alr1.pnew3;
set alr1.pnew2;
   z=log(pihat/(1-pihat))+((capsule-pihat)/(pihat*(1-pihat)));
   v=pihat*(1-pihat);
proc reg;
   model z=age racenew dcapsnew dpros_2 dpros_3 dpros_4 vol vol01
         gleason/selection=cp best=5;
   weight v;
run;
```

The above program was run, yielding the output shown below.

The LOGISTIC Procedure

Model Information

Data Set	ALR1.PNEW1
Response Variable	capsule
Number of Response Levels	2
Number of Observations	376
Link Function	Logit
Optimization Technique	Fisher's scoring

Response Profile

Ordered Value	capsule	Total Frequency
1	1	151
2	0	225

NOTE: 4 observations were deleted due to missing values for the response or explanatory variables.

Model Convergence Status

Convergence criterion (GCONV=1E-8) satisfied.

Model Fit Statistics

Criterion	Intercept Only	Intercept and Covariates
AIC	508.587	403.750
SC	512.517	443.046
-2 Log L	506.587	383.750

Testing Global Null Hypothesis: BETA=0

Test	Chi-Square	DF	Pr > ChiSq
Likelihood Ratio	122.8370	9	<.0001
Score	100.5981	9	<.0001
Wald	76.7375	9	<.0001

```
              Analysis of Maximum Likelihood Estimates

                              Standard
  Parameter    DF   Estimate    Error    Chi-Square    Pr > ChiSq

  Intercept    1    -7.4455    1.6104     21.3751       <.0001
  age          1    -0.0133    0.0196      0.4595       0.4978
  racenew      1    -0.3617    0.4471      0.6545       0.4185
  dcapsnew     1     0.6270    0.4453      1.9827       0.1591
  dpros_2      1     0.6450    0.3497      3.4012       0.0651
  dpros_3      1     1.3987    0.3702     14.2748       0.0002
  dpros_4      1     1.5189    0.4516     11.3105       0.0008
  vol          1     0.00112   0.0122      0.0085       0.9266
  vol01        1    -0.4824    0.4184      1.3292       0.2489
  gleason      1     1.1297    0.1614     48.9861       <.0001


                        The REG Procedure
                        Model: MODEL1
                    Dependent Variable: z

                          Weight: v

                     Analysis of Variance

                             Sum of        Mean
  Source            DF       Squares       Square    F Value    Pr > F

  Model              9      76.73847      8.52650     0.30      <.0001
  Error            366     376.11043      1.02762
  Corrected Total  375     452.84890


            Root MSE              1.01372    R-Square     0.1695
            Dependent Mean       -0.33302    Adj R-Sq     0.1490
            Coeff Var          -304.39803


                     Parameter Estimates

                       Parameter     Standard
  Variable      DF     Estimate       Error     t Value    Pr > |t|

  Intercept      1     -7.44559      1.63252     -4.56      <.0001
  age            1     -0.01331      0.01991     -0.67      0.5041
  racenew        1     -0.36168      0.45319     -0.80      0.4253
  dcapsnew       1      0.62705      0.45143      1.39      0.1657
  dpros_2        1      0.64499      0.35453      1.82      0.0697
  dpros_3        1      1.39875      0.37529      3.73      0.0002
  dpros_4        1      1.51887      0.45782      3.32      0.0010
  vol            1      0.00112      0.01237      0.09      0.9276
  vol01          1     -0.48239      0.42414     -1.14      0.2561
  gleason        1      1.12967      0.16362      6.90      <.0001
```

```
The REG Procedure
                                  Model: MODEL1
                              Dependent Variable: z

                              C(p) Selection Method

                                    Weight: v

Number in
  Model        C(p)    R-Square    Variables in Model

       5     4.7888     0.1631     dpros_2 dpros_3 dpros_4 vol01 gleason
       6     4.9858     0.1672     dcapsnew dpros_2 dpros_3 dpros_4 vol01 gleason
       6     6.3452     0.1641     dcapsnew dpros_2 dpros_3 dpros_4 vol gleason
       4     6.3615     0.1550     dpros_3 dpros_4 vol01 gleason
       6     6.3808     0.1641     racenew dpros_2 dpros_3 dpros_4 vol01 gleason
```

Note that the model coefficients from the LOGISTIC and REG procedures are the same for each covariate. The best subsets selection indicates that the model containing the indicator variables for DPROS, VOL01 and GLEASON is the best model on the basis of C_p. These are the same main effects that were identified using the purposeful selection strategy.

The covariates DPROS_2, DPROS_3, DPROS_4, VOL01 and GLEASON are identified as significant using all three modeling strategies. Stepwise and puposeful selection also identified PSA as an important main effect. PSA will be included in the final model. There is a significant interaction between DPROS and VOL01. The three interaction terms will be included in the final model. The final model for the prostatic cancer data is shown below.

Table. Logistic Regression Analysis of Factors Associated with CAPSULE=1 from the Prostatic Cancer Data Set, Final Model

Variable	Coefficient	Standard Error	Odds Ratio (95% CI)
Unilobar Nodule Left (vs. No Nodule)	1.663	0.5481	*
Unilobar Nodule Right (vs. No Nodule)	1.701	0.5945	*
Bilobar Nodule (vs. No Nodule)	2.627	0.7637	*
Tumor Volume $>1cm^3$ (vs. Tumor Vol=0)	0.442	0.5944	*
Prostatic Specific Antigen Value (mg/ml)	0.028	0.0098	1.33** (1.09, 1.61)
Total Gleason Score	1.020	0.1670	2.77*** (2.00, 3.85)
Unilobar Nodule Left*Volume	-1.743	0.7367	*
Unilobar Nodule Right*Volume	-0.248	0.7579	*
Bilobar Nodule*Volume	-2.040	0.9734	*

*Odds ratios not computed for interacting term. See separate table.
**Odds ratio for a 10-unit increase in PSA value.
***Odds ratio for a 1-unit increase in GLEASON score.

Table. Odds Ratios for Results of Digital Rectal Exam for Subjects with Tumor Volume Obtained from Ultrasound = 0 and Subjects with Tumor Volume Obtained from Ultrasound >0.

	Odds Ratio (95% Confidence Interval)	
	Volume=1	Volume=0
Unilobular Nodule Left (vs. No Nodule)	0.92 (0.35, 2.41)	5.28 (1.80, 15.44)
Unilobular Nodule Right (vs. No Nodule)	4.28 (1.68, 10.86)	5.48 (1.71, 17.57)
Bilobar Nodule (vs. No Nodule)	1.80 (0.55, 5.86)	13.83 (3.10, 61.79)

The confidence intervals suggest that there is considerable uncertainty in these estimates.

It would also be appropriate to report the odds ratio for VOL01=1 vs. VOL01=0 at each level of DPROS.

Table. Odds Ratios for VOL01=1 vs. VOL01=0 at each level of DPROS.

	DPROS=2	DPROS=3	DPROS=4
Odds Ratio for VOL01=1 VS. VOL01=0	0.27	1.21	0.20
95% Confidence Interval	(0.12, 0.63)	(0.40, 3.06)	(0.04, 0.91)

Chapter Five – Solutions

1. *As is the case in linear regression, effective use of diagnostic statistics depends on our ability to interpret and understand the values of the statistics. The purpose of this problem is to provide a few structured examples to examine the effect on the fitted logistic regression model and diagnostic statistics when data are moved away from the model (i.e., poorer fit), and also toward the model (i.e., better fit). Table 5.13 lists values of the independent variable, x, and seven different columns of the outcome variable, y, labeled "Model." All models fit in this problem use the given values of x for the covariate. Different models are fit using the seven different columns for the outcome variable. The data for the column labeled "Model 0" are constructed to represent a "typical" realization when the logistic regression model is correct. In the columns labeled "Model 1" to "Model 3" we have changed some of the y values away from the original model. Namely some cases with small values of x have had y changed from 0 to 1 and others with large values of x have had the y values changed from 1 to 0. For models labeled "Model-1" and "Model-2" we have moved the y values in the direction of the model. That is, we have changed y from 1 to 0 for some small values of x and have changed y from 0 to 1 for some large values of x. Fit the six logistic regression models for the data in columns "Model-2" to "Model 3." Compute for each fitted model the values of the leverage, h, the change in chi-square, ΔX^2, and the influence diagnostic, $\Delta \beta$.*

   Plot each of these versus the fitted values, predicted logistic probabilities. Compare the plots over the various models. Do the statistics pick out poorly fit and influential cases? How do the estimated coefficients change relative to Model 0? Fit "Model-i." What happens and why? Refer to the discussion in Section 4.5 on complete separation.

   Note that three plots are shown for each fitted model: (1) leverage versus predicted probabilities, (2) change in the Pearson chi-square versus the predicted probabilities and (3) Cook distance versus the predicted probabilities.

Model -2

```
. logit yn2 x

Iteration 0:   Log Likelihood =-13.460233
Iteration 1:   Log Likelihood =-5.8750882
Iteration 2:   Log Likelihood =-3.7918955
Iteration 3:   Log Likelihood =-2.7540781
Iteration 4:   Log Likelihood =-2.2922224
Iteration 5:   Log Likelihood =-2.1238781
Iteration 6:   Log Likelihood = -2.086162
Iteration 7:   Log Likelihood =-2.0833979
Iteration 8:   Log Likelihood =-2.0833778

Logit Estimates                              Number of obs =      20
                                             chi2(1)       =   22.75
                                             Prob > chi2   =  0.0000
Log Likelihood = -2.0833778                  Pseudo R2     =  0.8452

------------------------------------------------------------------------
     yn2 |    Coef.    Std. Err.      z     P>|z|    [95% Conf. Interval]
---------+--------------------------------------------------------------
       x |  3.234895   2.406123    1.344   0.179    -1.48102     7.95081
   _cons |  3.597932   2.880753    1.249   0.212   -2.048239    9.244103
------------------------------------------------------------------------

Note: 0 failures and 2 successes completely determined.
```

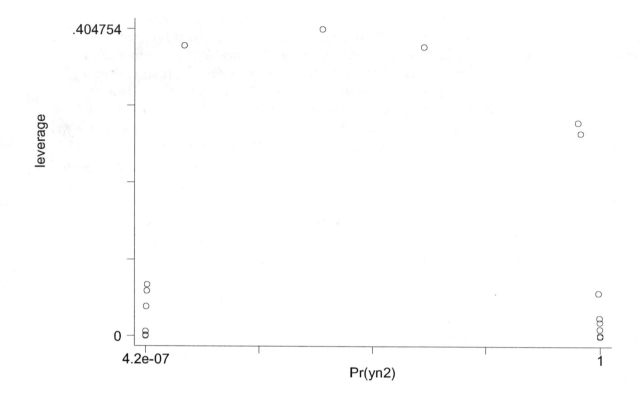

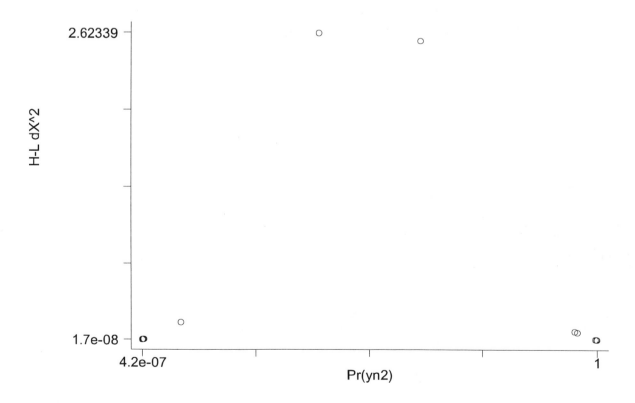

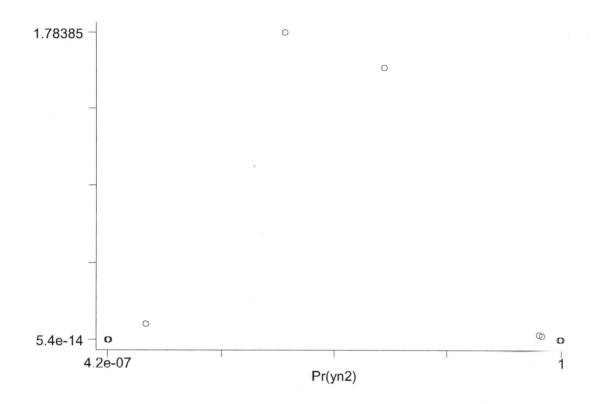

Model –1

```
. logit yn1 x

Iteration 0:  Log Likelihood =-13.762776
Iteration 1:  Log Likelihood =-7.3875599
Iteration 2:  Log Likelihood =-6.2532919
Iteration 3:  Log Likelihood =-5.9926508
Iteration 4:  Log Likelihood =-5.9705042
Iteration 5:  Log Likelihood =-5.9702782
Iteration 6:  Log Likelihood =-5.9702782

Logit Estimates                          Number of obs =      20
                                         chi2(1)       =   15.58
                                         Prob > chi2   = 0.0001
Log Likelihood = -5.9702782              Pseudo R2     = 0.5662


------------------------------------------------------------------------
    yn1 |    Coef.    Std. Err.      z      P>|z|    [ 95% Conf. Interval]
--------+---------------------------------------------------------------
      x |  1.091972   .4889585    2.233    0.026    .1336309    2.050313
  _cons |  .7625813   .7937227    0.961    0.337   -.7930867    2.318249
------------------------------------------------------------------------
```

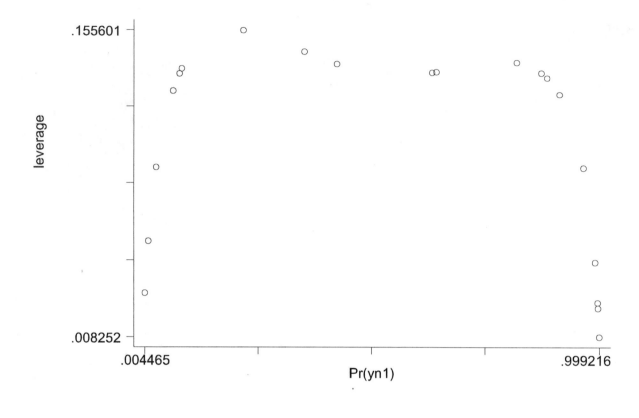

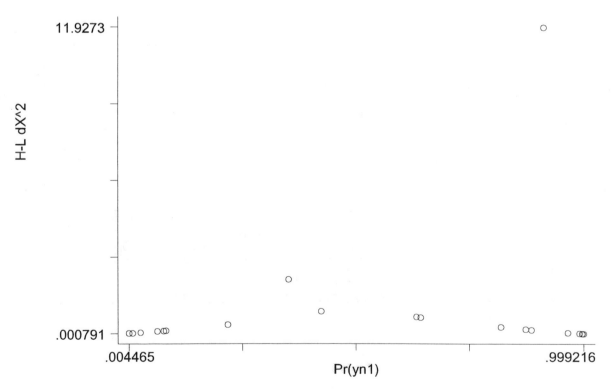

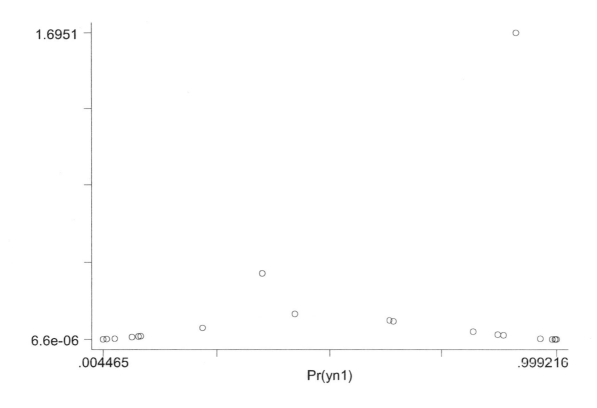

Model 0

```
.  logit y0 x

Iteration 0:   Log Likelihood =-13.862944
Iteration 1:   Log Likelihood =-8.9128019
Iteration 2:   Log Likelihood =-8.4153443
Iteration 3:   Log Likelihood =-8.3745648
Iteration 4:   Log Likelihood =-8.3741081
Iteration 5:   Log Likelihood =-8.3741081

Logit Estimates                                  Number of obs =       20
                                                 chi2(1)       =    10.98
                                                 Prob > chi2   =   0.0009
Log Likelihood = -8.3741081                      Pseudo R2     =   0.3959

------------------------------------------------------------------------
     y0 |      Coef.   Std. Err.       z     P>|z|    [ 95% Conf. Interval]
--------+---------------------------------------------------------------
      x |   .6928009   .3009826     2.302    0.021     .1028858    1.282716
  _cons |    .139359   .6140216     0.227    0.820    -1.064101    1.342819
------------------------------------------------------------------------
```

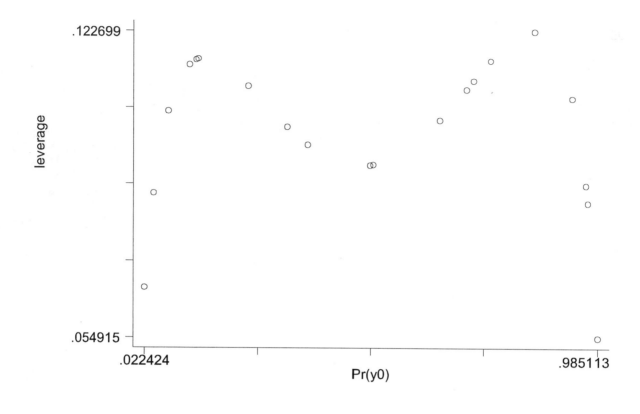

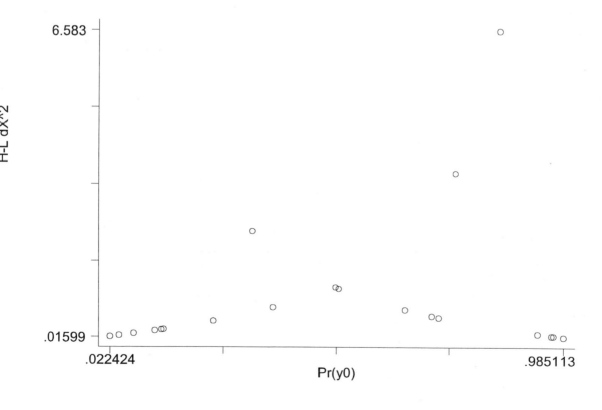

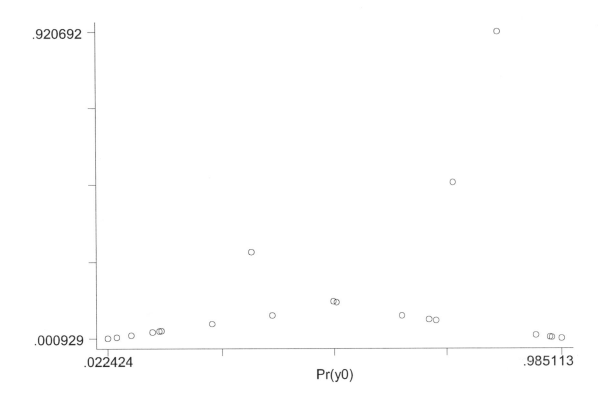

Model 1

```
.  logit y1 x

Iteration 0:   Log Likelihood =-13.762776
Iteration 1:   Log Likelihood =-9.5990874
Iteration 2:   Log Likelihood =-9.2665939
Iteration 3:   Log Likelihood =-9.2499427
Iteration 4:   Log Likelihood =-9.2498782

Logit Estimates                              Number of obs =      20
                                             chi2(1)       =    9.03
                                             Prob > chi2   = 0.0027
Log Likelihood = -9.2498782                  Pseudo R2     = 0.3279


------------------------------------------------------------------------
     y1  |    Coef.    Std. Err.      z     P>|z|    [ 95% Conf. Interval]
---------+--------------------------------------------------------------
      x  |  .5833248   .2589968     2.252   0.024    .0757005    1.090949
   _cons |  .4423584   .5974215     0.740   0.459   -.7285661    1.613283
------------------------------------------------------------------------
```

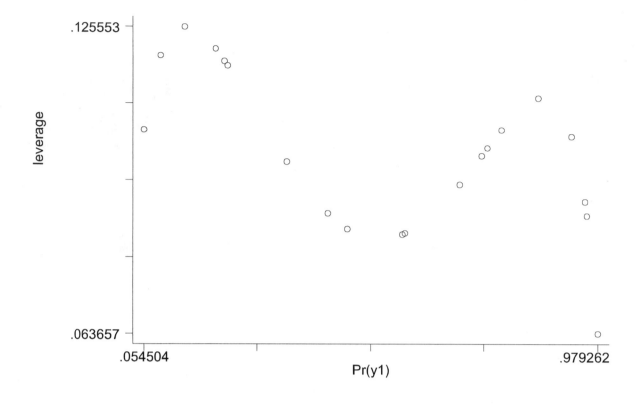

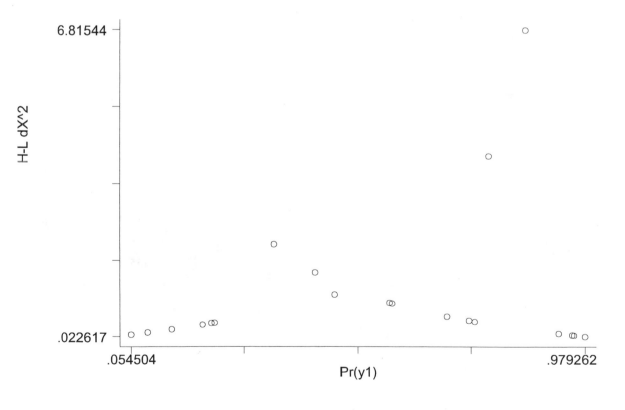

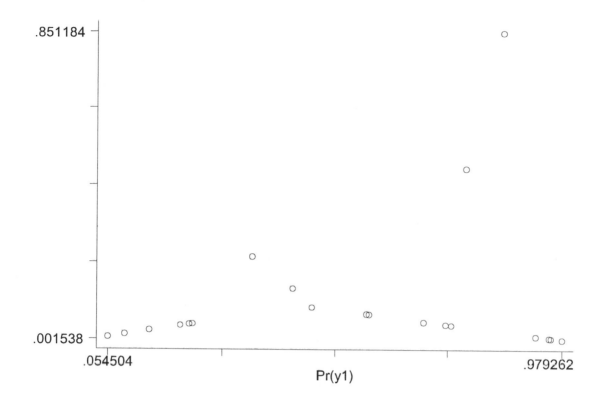

Model 2

```
. logit y2 x

Iteration 0:   Log Likelihood =-13.460233
Iteration 1:   Log Likelihood =-11.064063
Iteration 2:   Log Likelihood =-10.973204
Iteration 3:   Log Likelihood =-10.972157
Iteration 4:   Log Likelihood =-10.972157

Logit Estimates                              Number of obs =      20
                                             chi2(1)       =    4.98
                                             Prob > chi2   = 0.0257
Log Likelihood = -10.972157                  Pseudo R2     = 0.1848

------------------------------------------------------------------------
     y2 |     Coef.    Std. Err.      z      P>|z|    [95% Conf. Interval]
--------+---------------------------------------------------------------
      x |   .379892     .19865      1.912   0.056    -.0094549    .7692389
  _cons |  .5929312    .5439606     1.090   0.276    -.473212    1.659074
------------------------------------------------------------------------
```

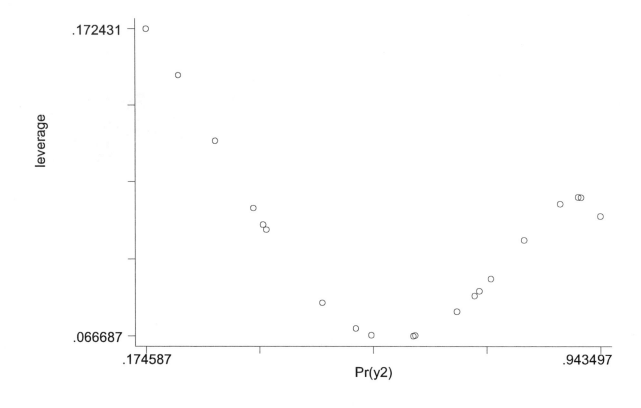

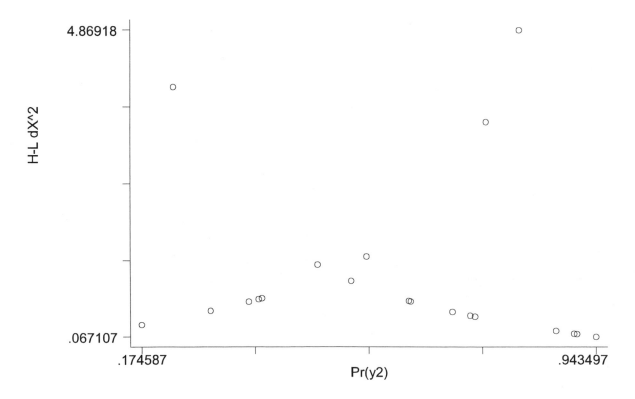

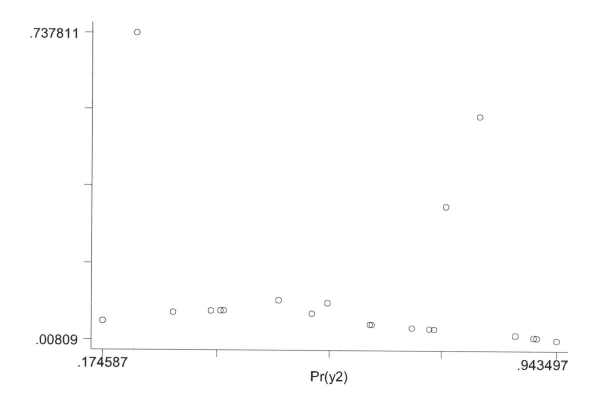

Model 3

```
. logit y3 x

Iteration 0:  Log Likelihood =-13.762776
Iteration 1:  Log Likelihood =-12.725204
Iteration 2:  Log Likelihood =-12.718277
Iteration 3:  Log Likelihood =-12.718274

Logit Estimates                          Number of obs =      20
                                         chi2(1)       =    2.09
                                         Prob > chi2   = 0.1484
Log Likelihood = -12.718274              Pseudo R2     = 0.0759


------------------------------------------------------------------------
     y3 |    Coef.    Std. Err.      z      P>|z|     [95% Conf. Interval]
--------+---------------------------------------------------------------
      x |   .218659    .160551    1.362    0.173    -.0960152    .5333332
  _cons |   .254074    .4775628   0.532    0.595    -.6819319    1.19008
------------------------------------------------------------------------
```

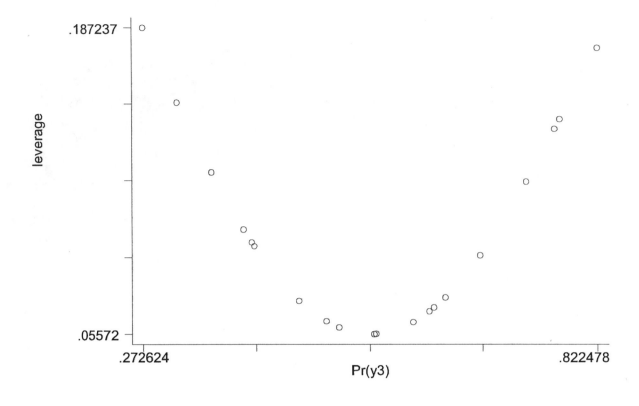

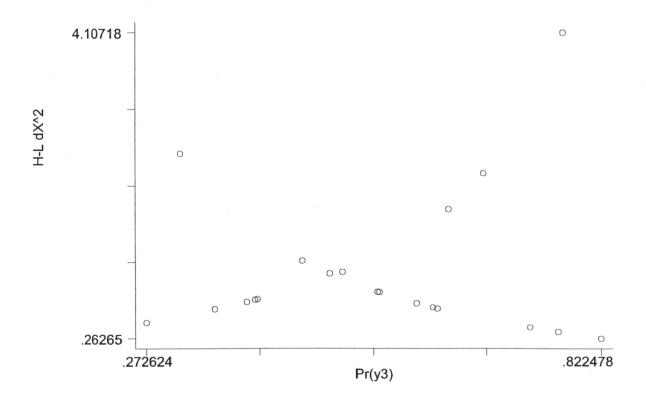

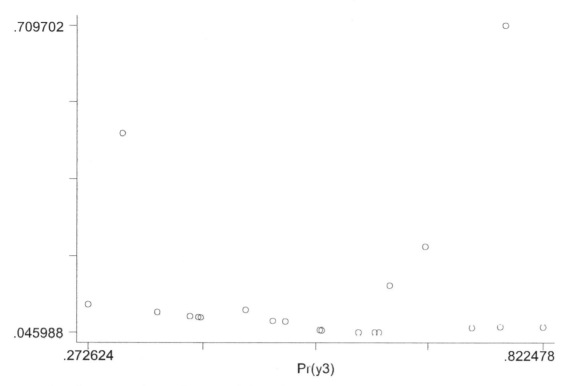

The plots over the various models indicate points that are poorly fit and influential. For example, in Model –2, the graph of $\Delta\beta$ identifies observations 8 and 9 ($x = -1.25$ and $x = -0.97$, respectively) as being influential. In Model –1, the graph of $\Delta\beta$ identifies observation 15 ($x = 1.45$) as being influential. This observation does not have the outcome present, despite having a relatively large value of x. Models 0, 1, 2 and 3 do not appear to have subjects that are as influential as the two models just discussed. The output from the logistic regression of each y on x, shows that in Models –1 and –2 the coefficient for x becomes larger, the further the model moves toward from the "typical model", Model 0. As the values of y move away from the typical model (Models 1, 2 and 3), the coefficient for x decreases.

Fit "Model –i". What happens and why?

Model –i

```
. logistic yi x
outcome = x>-.97 predicts data perfectly
r(2000);
```

Model –i cannot be fit. There is complete separation. This means that since all subjects with a value of x<-0.2 do not have the outcome, while all subjects with a value ≥-0.19 do have the outcome, we know with certainty the value of the outcome variable, if we know the value of x. Since there is no overlap in the distribution of the covariates in the model, maximum likelihood estimates do not exist.

2.   *In the exercises in Chapter 4, Problem 3, multivariable models for the ICU study were formed. Assess the fit of the model that you feel was best among those considered. This assessment should include an overall assessment of fit and use of the diagnostic statistics. Does the model fit? Are there any particular subjects, or covariate patterns, which seem to be poorly fit or overly influential? If so, how would you propose to deal with them?*

The best multivariable model for the ICU data from Chapter 4, Problem 3 is shown below.

```
. logistic sta loc2 typ age can sys agesys

Logit Estimates                              Number of obs =      200
                                             chi2(6)       =    69.32
                                             Prob > chi2   =   0.0000
Log Likelihood = -65.420861                  Pseudo R2     =   0.3463


--------------------------------------------------------------------------
    sta | Odds Ratio   Std. Err.       z     P>|z|    [ 95% Conf. Interval]
--------+-----------------------------------------------------------------
   loc2 |   45.39315    42.68579     4.057    0.000    7.187146    286.6977
    typ |     18.886    17.34661     3.199    0.001    3.121185    114.2774
    age |   1.187805    .0787806     2.595    0.009    1.043013    1.352698
    can |   9.275526    7.801173     2.648    0.008     1.78417    48.22151
    sys |   1.055576    .0329379     1.733    0.083    .9929533    1.122148
 agesys |   .9989684     .000478    -2.157    0.031    .9980319    .9999056
--------------------------------------------------------------------------
```

Use the Hosmer-Lemeshow test to evaluate the overall fit of the model.

```
. lfit, group (10) table

Logistic model for sta, goodness-of-fit test
 (Table collapsed on quantiles of estimated probabilities)

_Group      _Prob     _Obs_1     _Exp_1     _Obs_0     _Exp_0      _Total
     1     0.0091          0        0.1         20       19.9          20
     2     0.0234          0        0.3         20       19.7          20
     3     0.0513          1        0.8         19       19.2          20
     4     0.0793          2        1.3         18       18.7          20
     5     0.1187          1        2.0         19       18.0          20
     6     0.1411          2        2.6         18       17.4          20
     7     0.2019          4        3.5         16       16.5          20
     8     0.2788          6        4.9         14       15.1          20
     9     0.5229          8        7.7         12       12.3          20
    10     0.9771         16       16.8          4        3.2          20

           number of observations =        200
                  number of groups =         10
        Hosmer-Lemeshow chi2(8) =       2.38
                   Prob > chi2 =     0.9669
```

The test indicates that the fit of the model is adequate.

Use the area under the ROC Curve to assess the model's ability to discriminate between those subjects with the outcome versus those without the outcome.

```
. lroc

Logistic model for sta

number of observations =      200
area under ROC curve   =   0.8689
```

The area under the ROC curve, 0.8689, indicates excellent discrimination.

Use diagnostic statistics to identify poorly fit or overly influential subjects.

`. graph dx2 phat`

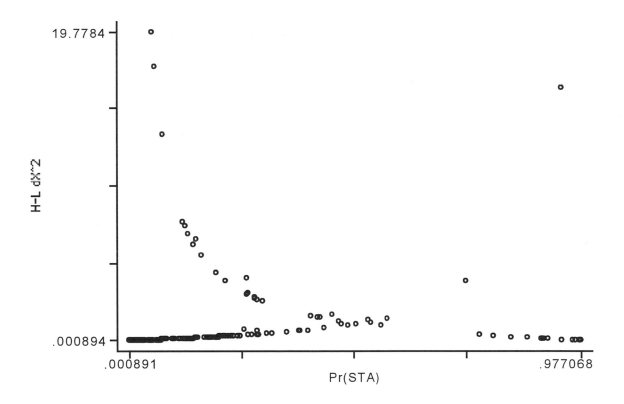

. graph ddeviance phat

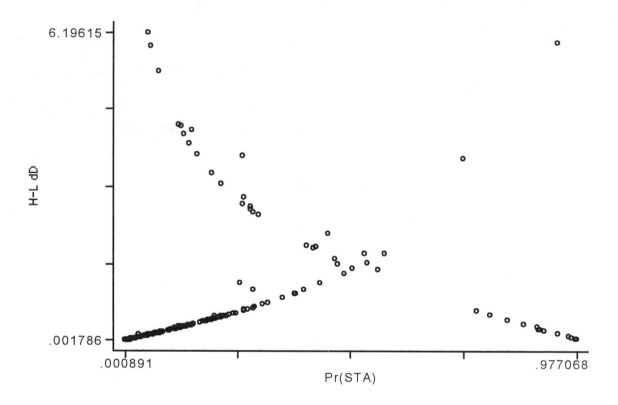

. graph dbeta phat

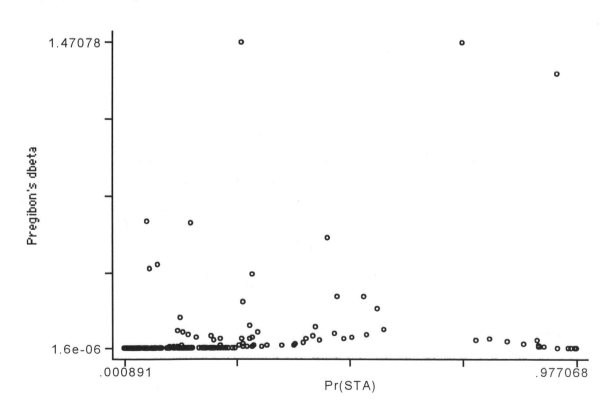

. graph dx2 phat [w= dbeta]

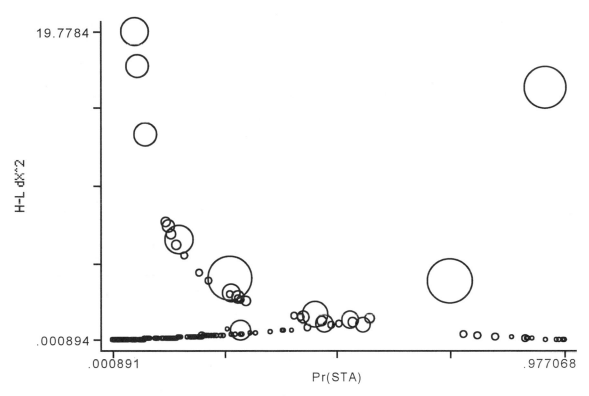

Pr(STA)

Covariate patterns with large leverage, residual and influence diagnostic statistics are shown in the table below.

ID	AGE	SYS	CAN	TYP	LOC2	STA	π	h	ΔX^2	$\Delta\beta$	Large
881	89	190	0	1	1	0	0.73	0.28	3.82	1.47	$\Delta\beta$
208	70	168	0	0	1	1	0.26	0.27	3.98	1.47	$\Delta\beta$
84	59	48	0	1	1	0	0.94	0.08	16.16	1.33	$\Delta\beta, \Delta X^2$
127	19	140	0	1	0	1	0.06	0.02	17.53	0.38	ΔX^2
285	40	86	0	1	0	1	0.05	0.03	19.78	0.61	ΔX^2
380	20	148	0	1	0	1	0.07	0.03	13.16	0.40	ΔX^2
645	36	224	0	1	0	0	0.28	0.37	0.61	0.36	h

The effect each of these covariate patterns has on the model is shown in the following table. The values and percent changes in estimated coefficients after deleting the cases with the largest values of the diagnostic statistics are shown.

Variable	ALL OBS	drop 881	% change	drop 208	% change	drop 84	% change	drop 127	% change
AGE	.172	.139	-19.3	.178	3.1	.163	-5.3	.178	3.5
SYS	.054	.038	-28.8	.055	2.3	.045	-17.1	.053	-1.9
AGESYS	-.001	-0.0007	-27.5	-.001	3.9	-.0009	-5.1	-.001	0
CAN	2.227	2.304	3.4	2.632	18.1	2.341	-5.1	2.371	6.5
TYP	2.938	3.160	7.5	3.849	30.9	3.131	-6.5	2.986	1.6
LOC2	3.815	4.708	23.3	3.256	14.4	5.074	-33.0	3.892	2.0
CONS	-13.877	-12.255	-11.7	-14.959	7.8	-12.738	-8.2	-14.345	-3.4

drop 285	% change	drop 380	% change	drop 645	% change
.220	37.8	.166	-3.5	.204	18.6
.076	41.1	.047	-13.0	.072	32.8
-.001	30.2	-0.0009	-10.0	-.001	23.9
2.389	7.2	2.353	5.7	2.224	-.1
2.989	-1.6	2.978	1.4	2.960	.7
3.942	3.0	3.883	1.8	3.831	.4
-17.380	25.0	-13.513	2.6	-16.204	16.7

Observations 881, 285 and 645 seem to have the most extreme impact on the model's coefficients. At this point, it would be advisable to check for coding errors for those subjects and consult with a subject matter specialist on the clinical plausibility of the data.

3. *Repeat Exercise 2 for the models developed for the low birth weight data and the prostate cancer data in Chapter 4, Exercise 4.*

Low Birthweight Data

The best multivariable model for the Low Birthweight data from Chapter 4, Exercise 4 is shown below.

```
. logistic low smoke ptld age ht lwt race_2 race_3

Logit Estimates                              Number of obs =      189
                                             chi2(7)       =    35.52
                                             Prob > chi2   = 0.0000
Log Likelihood = -99.575705                  Pseudo R2     = 0.1514

------------------------------------------------------------------------
    low | Odds Ratio  Std. Err.      z    P>|z|   [95% Conf. Interval]
--------+---------------------------------------------------------------
  smoke |  2.359223   .9550055    2.120   0.034   1.067101   5.215939
   ptld |  3.796083   1.737005    2.915   0.004   1.548262   9.307368
    age |  .958118    .0359947   -1.139   0.255   .8901045   1.031329
     ht |  5.700254   4.007922    2.475   0.013   1.436824   22.61439
    lwt |  .9846826   .0069363   -2.191   0.028   .9711812   .9983717
 race_2 |  3.217009   1.713329    2.194   0.028   1.132702   9.136693
 race_3 |  2.258318   .9998732    1.840   0.066   .9482319   5.378432
------------------------------------------------------------------------
```

Use the Hosmer-Lemeshow test to evaluate the overall fit of the model.

```
. lfit, group (10) table

Logistic model for low, goodness-of-fit test
 (Table collapsed on quantiles of estimated probabilities)

_Group      _Prob     _Obs_1     _Exp_1     _Obs_0     _Exp_0     _Total
     1     0.0832         0        1.2        19       17.8        19
     2     0.1295         3        2.0        16       17.0        19
     3     0.1881         3        3.0        16       16.0        19
     4     0.2378         5        4.1        14       14.9        19
     5     0.2752         4        4.9        15       14.1        19
     6     0.3108         6        5.6        13       13.4        19
     7     0.3704         8        6.7        12       13.3        20
     8     0.5046         4        7.8        14       10.2        18
     9     0.6071        14       10.8         5        8.2        19
    10     0.8932        12       12.9         6        5.1        18

             number of observations  =        189
                   number of groups  =         10
          Hosmer-Lemeshow chi2(8)    =       8.31
                    Prob > chi2      =     0.4040
```

The test indicates that the fit of the model is adequate.

Use the area under the ROC Curve to assess the model's ability to discriminate between those subjects with the outcome versus those without the outcome.

```
. lroc

Logistic model for low

number of observations  =        189
area under ROC curve    =     0.7499
```

The area under the ROC curve, 0.7499, indicates acceptable discrimination.

Use diagnostic statistics to identify poorly fit or overly influential subjects.

Covariate patterns with large leverage, residual and influence diagnostic statistics are shown in the table below.

ID	SMOKE	PTLD	AGE	HT	LWT	RACE	LOW	π	h	ΔX^2	$\Delta\beta$
10	0	0	29	0	130	1	1	0.09	0.017	10.38	0.184
11	1	0	34	1	187	2	1	0.59	0.193	0.87	0.209
28	0	0	21	0	200	2	1	0.13	0.046	6.95	0.332
36	0	0	24	0	138	1	1	0.10	0.016	9.46	0.140
77	1	0	26	0	190	1	1	0.09	0.028	9.88	0.282
98	0	0	22	1	95	3	0	0.75	0.119	3.31	0.446
119	1	1	35	0	121	2	0	0.72	0.115	2.84	0.369

The effect each of these covariate patterns has on the model is shown in the following table. The values and percent changes in estimated coefficients after deleting the cases with the largest values of the diagnostic statistics are shown.

Variable	ALL OBS	drop 10	% change	drop 11	% change	drop 28	% change	drop 36	% change
SMOKE	0.858	.966	12.6	.803	-6.4	.907	5.7	.963	12.2
PTLD	1.334	1.383	3.6	1.373	2.9	1.345	0.8	1.362	2.1
AGE	-.043	-.051	18.6	-.051	18.6	-.040	-7.0	-.044	2.3
HT	1.741	1.786	2.6	1.559	-10.5	1.893	8.7	1.799	3.3
LWT	-.015	-.015	0	-.016	6.7	-.018	0.2	-.016	6.7
RACE_2	1.168	1.258	7.7	1.057	-9.5	1.057	-9.5	1.273	9.0
RACE_3	.815	.932	14.4	.771	-5.4	.823	1.0	.932	14.4
CONS	.925	.964	4.2	1.194	29.1	1.181	27.7	.875	-5.4

drop 77	% change	drop 98	% change	drop 119	% change
.798	-7.0	0.829	-3.4	.920	7.2
1.387	4.0	1.317	-1.3	1.423	6.7
-.045	4.7	-.041	-4.7	-.029	-32.6
1.892	8.7	2.190	25.8	1.737	-0.2
-.019	2.7	-.018	20.0	-.016	6.7
1.261	8.0	1.185	1.5	1.380	18.2
0.823	1.0	.857	5.2	.858	5.3
1.334	44.2	1.186	28.2	.669	-27.7

Observations 98 and 119 seem to have the most extreme impact on the model's coefficients. At this point, it would be advisable to check for coding errors for those subjects and consult with a subject matter specialist on the clinical plausibility of the data.

Prostatic Cancer Data

The best multivariable model for the prostatic cancer data from Chapter 4, Exercise 3 is shown below.

```
. logistic capsule  dpros_2 dpros_3 dpros_4 gleason psa vol01 d2vol d3vol d4vol

Logit Estimates                              Number of obs =      380
                                             chi2(9)       = 145.21
                                             Prob > chi2   = 0.0000
Log Likelihood = -183.53939                  Pseudo R2     = 0.2835


-------------------------------------------------------------------------
capsule | Odds Ratio  Std. Err.      z     P>|z|    [ 95% Conf. Interval]
--------+----------------------------------------------------------------
dpros_2 |  5.275048   2.891255     3.034   0.002     1.8017    15.44437
dpros_3 |  5.481237   3.258331     2.862   0.004    1.709544   17.57425
dpros_4 |  13.83012   10.56266     3.439   0.001    3.095471   61.79099
gleason |  2.772111    .462972     6.105   0.000    1.998255   3.845655
    psa |  1.028601   .0100604     2.883   0.004    1.009071    1.04851
  vol01 |  1.555232   .9243524     0.743   0.457     .485158    4.98548
  d2vol |  .1749237   .1288683    -2.366   0.018    .0412818   .7412055
  d3vol |  .7806982   .5917255    -0.327   0.744     .176734   3.448627
  d4vol |  .1300411   .1265868    -2.096   0.036     .019297   .8763378
-------------------------------------------------------------------------
```

Use the Hosmer-Lemeshow test to evaluate the overall fit of the model.

```
. lfit, group (10) table

Logistic model for capsule, goodness-of-fit test
 (Table collapsed on quantiles of estimated probabilities)

_Group    _Prob    _Obs_1    _Exp_1    _Obs_0    _Exp_0    _Total
     1    0.0621         2       1.7        36      36.3        38
     2    0.1304         6       4.1        32      33.9        38
     3    0.1583         5       5.4        33      32.6        38
     4    0.2618         4       7.8        34      30.2        38
     5    0.3503        10      12.1        28      25.9        38
     6    0.4004        15      14.2        23      23.8        38
     7    0.5921        22      19.2        16      18.8        38
     8    0.6813        26      24.3        12      13.7        38
     9    0.8560        29      28.8         9       9.2        38
    10    0.9962        34      35.4         4       2.6        38

          number of observations =        380
                 number of groups =         10
        Hosmer-Lemeshow chi2(8) =        6.01
                   Prob > chi2 =        0.6456
```

The test indicates that the fit of the model is adequate.

Use the area under the ROC Curve to assess the model's ability to discriminate between those subjects with the outcome versus those without the outcome.

```
. lroc

Logistic model for capsule

number of observations =        380
area under ROC curve    =     0.8388
```

The area under the ROC curve, 0.8388, indicates excellent discrimination.

Use diagnostic statistics to identify poorly fit or overly influential subjects.

Covariate patterns with larger leverage, residual and influence diagnostic statistics are shown in the table below. Plots of the diagnostic statistics do not reveal any covariate patterns that appear to be particularly poorly fit or overly influential.

ID	DPROS	GLEASON	PSA	VOL01	CAPSULE	π	h	ΔX^2	$\Delta\beta$	Large
7	4	7	31.9	0	0	.89	0.04	8.40	0.33	$\Delta\beta, \Delta X^2$
8	4	7	66.7	1	0	0.81	0.07	4.68	0.33	$\Delta\beta$
33	4	8	11	0	0	0.93	0.03	12.84	0.41	$\Delta\beta, \Delta X^2$
89	4	9	17.1	1	0	0.89	0.04	8.59	0.32	$\Delta\beta, \Delta X^2$
278	2	5	8.4	1	1	0.05	0.01	17.9	0.16	ΔX^2
312	4	5	9.9	1	1	0.10	0.03	8.98	0.25	ΔX^2
274	1	6	11.2	0	1	0.11	0.02	8.68	0.19	ΔX^2
192	4	6	61.6	1	0	0.58	0.11	1.53	0.19	h
76	1	7	64.3	0	0	0.59	0.10	1.62	0.18	h

The effect each of these covariate patterns has on the model is shown in the following table. The values and percent changes in estimated coefficients after deleting the cases with the largest values of the diagnostic statistics are shown.

Variable	ALL OBS	drop 7	% change	drop 8	% change	drop 33	% change	drop 89	% change
DPROS_2	1.663	1.671	0.5	1.688	1.5	1.665	0.1	1.672	0.5
DPROS_3	1.701	1.708	0.4	1.722	1.2	1.703	0.1	1.709	0.5
DPROS_4	2.627	2.999	14.2	2.614	-0.5	3.026	15.2	2.641	0.5
GLEASON	1.020	1.029	0.9	1.009	-1.1	1.065	4.4	1.082	6.1
PSA	.028	.029	3.6	0.032	14.3	.027	-3.6	.027	-3.6
VOL01	.442	.445	0.7	0.445	0.7	.452	2.2	.457	3.4
D2VOL	-1.743	-1.754	0.6	-1.765	1.3	-1.761	1.0	-1.7733	1.7
D3VOL	-.248	-.249	0.4	-.246	-0.9	-.257	3.6	-.260	4.8
D4VOL	-2.040	-2.416	18.4	-1.840	-9.8	-2.448	20.0	-1.862	-8.7
CONS	-8.573	-8.648	0.9	-8.575	0	-8.854	3.3	-8.973	4.7

drop 278	% change	drop 312	% change	drop 274	% change	drop 192	% change	drop 76	% change
1.673	0.6	1.673	0.6	1.874	12.7	1.677	0.8	1.544	-7.2
1.710	0.5	1.710	0.5	1.912	12.4	1.713	0.7	1.580	-7.1
2.634	0.3	2.633	0.2	2.835	7.9	2.615	-0.4	2.481	-5.6
1.063	4.2	1.060	3.9	1.035	1.5	.995	-2.5	1.012	-0.8
.028	0	.028	0	.028	0	.031	10.7	.031	10.7
.453	2.5	.452	2.3	.652	47.5	.439	-0.7	.307	-30.5
-1.865	7.0	-1.767	1.4	-1.959	12.4	-1.749	0.3	-1.622	-6.9
-.256	3.2	-.256	3.2	-.457	84.3	-.243	-2.0	-.109	-56.0
-2.059	0.9	-2.268	11.2	-2.252	10.4	-1.884	-7.4	-1.899	-6.9
-8.864	3.4	-8.846	3.2	-8.883	3.6	-8.462	-1.3	-8.439	-1.6

Observations 274 and 76 seem to have the most extreme impact on the model's coefficients. At this point, it would be advisable to check for coding errors for those subjects and consult with a subject matter specialist on the clinical plausibility of the data.

4. *Fit the final model for the UIS shown in Table 5.10 and obtain the estimated covariance matrix of the estimated coefficients.*

```
. logit dfree age  ndrgfp1 ndrgfp2 ivhx_2 ivhx_3 race treat site agend1 racesite

Iteration 0:   Log Likelihood = 326.86446

Iteration 1:   Log Likelihood =-300.06724
Iteration 2:   Log Likelihood =-298.98837
Iteration 3:   Log Likelihood =-298.98146
Iteration 4:   Log Likelihood =-298.98146

Logit Estimates                              Number of obs =      575
                                             chi2(10)      =    55.77
                                             Prob > chi2   = 0.0000
Log Likelihood = -298.98146                  Pseudo R2     = 0.0853


------------------------------------------------------------------------------
  dfree |     Coef.   Std. Err.      z     P>|z|     [95% Conf. Interval]
--------+---------------------------------------------------------------------
    age |   .1166385   .0288749     4.039   0.000     .0600446    .1732323
 ndrgfp1 |  1.669035    .407152     4.099   0.000     .871032    2.467038
 ndrgfp2 |  .4336886   .1169052     3.710   0.000     .2045586    .6628185
  ivhx_2 |  -.6346307   .2987192    -2.125   0.034    -1.220109   -.0491518
  ivhx_3 |  -.7049475   .2615805    -2.695   0.007    -1.217636   -.1922591
    race |   .6841068   .2641355     2.590   0.010     .1664107    1.201803
   treat |   .4349255   .2037596     2.135   0.033     .035564     .834287
    site |   .516201    .2548881     2.025   0.043     .0166295    1.015773
  agend1 |  -.0152697   .0060268    -2.534   0.011    -.0270819   -.0034575
 racesite | -1.429457   .5297806    -2.698   0.007    -2.467808    .3911062
   _cons |  -6.843864   1.219316    -5.613   0.000    -9.23368    -4.454048
------------------------------------------------------------------------------
```

```
. corr, _coef cov

        |     age  ndrgfp1  ndrgfp2   ivhx_2   ivhx_3     race    treat
--------+---------------------------------------------------------------------
    age|  .000834
 ndrgfp1|  .007339  .165773
 ndrgfp2|  .00113    .0424   .013667
  ivhx_2| -.000941  .007654  .000744  .089233
  ivhx_3| -.00145   .009334  .002499  .033625  .068424
    race|  .000317  .007374  .001309  .001977  .012888  .069768
   treat|  .000079 -.002815 -.000623  -.00399 -.000252 -.000734  .041518
    site| -.000057   .00006  .000819 -.009424  .007779  .025822  .003688
  agend1| -.000135 -.001694 -.000211 -.000121 -.000012 -.000116  .000042
 racesite| -.000668 -.002053 -.000402  .016725 -.000973 -.067752 -.007411
   _cons|  -.03165 -.418563 -.092098 -.004046 -.008189 -.046354 -.023178

        |    site   agend1  racesite    _cons
--------+------------------------------------
    site|  .064968
  agend1|  .000032  .000036
 racesite| -.062545  -.00002  .280667
   _cons| -.025728  .005303  .055338  1.48673
```

5. *Use the method of logit differences to estimate the odds ratio for site A versus B within racial groups.*

Site B (SITE= 1) vs. Site A (SITE= 0), Race = White (RACE=0)

$$\ln(OR) = \beta_1(SITE) + \beta_2(RACE) + \beta_3(RACESITE) -$$
$$\left[\beta_1(SITE) + \beta_2(RACE) + \beta_3(RACESITE)\right]$$
$$\ln(OR) = (0.516201)(1) = 0.515201$$

$$OR = \exp\left[\ln(OR)\right] = 1.68$$

Site B (SITE=1) vs. Site A (SITE=0), Race = White (RACE=0)

$$\ln(OR) = \beta_1(SITE) + \beta_2(RACE) + \beta_3(RACESITE) -$$
$$\left[\beta_1(SITE) + \beta_2(RACE) + \beta_3(RACESITE)\right]$$

$$\ln(OR) = (0.516201)(1) + (-1.429457)(1) = -0.913256$$

$$OR = \exp\left[\ln(OR)\right] = 0.40$$

6. *As a more complicated exercise in using the method of logit differences used in Section 5.5, estimate the odds ratio for a 25 year old white subject with 1 previous drug treatment, no history of IV drug use and on the longer treatment at site A compared to a 30 year old non-white subject with 5 previous drug treatments, a recent IV drug user and on the shorter treatment at site B. We make no claim that this is a clinically useful comparison. The purpose of the exercise is to illustrate the general applicability of the method of logit differences.*

AGE=25, NDRUGTX–1, RACE=0, IVIIX=1, TREAT=1, SITE=0 (IVHX_2=0, IVHX_3–0, NDRGFP1=5, NDRGFP2=-8.047)

vs.

AGE=30, NDRUGTX=5, RACE=1, IVHX=3, TREAT=0, SITE=1 (IVHX_2=0, IVHX_3=1, NDRGFP1=1.667, NDRGFP2=-0.8514)

$$\ln(OR) = \beta_1(AGE) + \beta_2(NDRGFP1) + \beta_3(NDRGFP2) + \beta_4(IVHX\_2) + \beta_5(IVHX\_3)$$
$$+ \beta_6(RACE) + \beta_7(TREAT) + \beta_8(SITE) + \beta_9(AGEND1) + \beta_{10}(RACESITE) -$$
$$\left[\begin{matrix} \beta_1(AGE) + \beta_2(NDRGFP1) + \beta_3(NDRGFP2) + \beta_4(IVHX\_2) + \beta_5(IVHX\_3) \\ + \beta_6(RACE) + \beta_7(TREAT) + \beta_8(SITE) + \beta_9(AGEND1) + \beta_{10}(RACESITE) \end{matrix} \right]$$
$$\ln(OR) = (0.1166385)(25) + (1.669035)(5) + (0.4336886)(-8.047) +$$
$$+ (0.4349255)(1) + (-0.0152697)(125) -$$
$$\left[\begin{matrix} (0.1166385)(30) + (1.669035)(1.6667) + (0.4336886)(-0.8514) + (-0.7049475)(1) \\ + (0.6841068)(1) + (0.516201)(1) + (-0.0152697)(50) + (-1.429457)(1) \end{matrix} \right]$$

$$\ln(OR) = 2.083402511$$

$$OR = \exp[\ln(OR)] = 8.03$$

7. *Obtain 95% confidence intervals for the estimates in Exercises 5 and 6.*

Site B (SITE=1) vs. Site A (SITE=0), Race = White (RACE=0)

Calculate the variance

$$Var\{\ln[OR]\} = (1)^2 * Var(\beta_1) = 0.064968$$

Once the variance is known, the standard error can be obtained by taking its square root and the confidence interval can be formed in the usual manner.

The odds ratio for Site B vs. Site A for white subjects calculated in Exercise 5 was 1.68. Using the variance calculated above, the 95% confidence interval for this odds ratio is (1.02, 2.76).

Site B (SITE=1) vs. Site A (SITE=0), Race = Other (RACE=1)

Calculate the variance, x=Race, f₁=Site B, f₀=Site A

$$Var\{\ln[OR]\} = (f_1 - f_0)^2 * Var(\beta_1) + [x(f_1 - f_0)]^2 * Var(\beta_3) + 2x(f_1 - f_0) * Cov(\beta_1\beta_3)$$

$$Var\{\ln[OR]\} = (1)^2 * (0.064968) + [1(1)]^2 * (0.28067) + 2(1)(1) * (-0.062545) = 0.22055$$

```
. lincom _b[site] + _b[ racxssit]

 ( 1)   site + racxssit = 0.0

------------------------------------------------------------------------
      dfree |      Coef.    Std. Err.      z     P>|z|    [95% Conf. Interval]
------------+-----------------------------------------------------------
        (1) |   -.913256    .4696229    -1.94   0.052    -1.8337      .007188
------------------------------------------------------------------------

. lincom _b[site] + _b[ racxssit],or

 ( 1)   site + racxssit = 0.0

------------------------------------------------------------------------
      dfree | Odds Ratio   Std. Err.      z     P>|z|    [95% Conf. Interval]
------------+-----------------------------------------------------------
        (1) |   .4012157    .1884201    -1.94   0.052    .1598211    1.007214
------------------------------------------------------------------------
```

Once the variance is known, the standard error can be obtained by taking its square root and the confidence interval can be formed in the usual manner.

The odds ratio for Site B vs. Site A for non-white subjects calculated in Exercise 5 was 0.40. Using the variance calculated above, the 95% confidence interval for this odds ratio is (0.16, 1.01).

AGE=25, NDRUGTX=1, RACE=0, IVHX=1, TREAT=1, SITE=0 (IVHX_2=0, IVHX_3=0, NDRGFP1=5, NDRGFP2=-8.047)

vs.

AGE=30, NDRUGTX=5, RACE=1, IVHX=3, TREAT=0, SITE=1 (IVHX_2=0, IVHX_3=1, NDRGFP1=1.667, NDRGFP2=-0.8514)

$$Var\{\ln[OR]\} = (f_1 - f_0)^2 * Var(\beta_1) + (g_1 - g_0)^2 * Var(\beta_2) + \cdots + (o_1 - o_0)^2 * Var(\beta_{10})$$
$$+ (f_1 - f_0)(g_1 - g_0)Cov(\beta_1\beta_2) + \cdots + (f_1 - f_0)(o_1 - o_0)Cov(\beta_1\beta_{10})$$
$$+ \cdots + (n_1 - n_0)(o_1 - o_0)Cov(\beta_9\beta_{10})$$

In the above equation, $(f_1 - f_0)$ represents the difference in AGE between the two logits, $(g_1 - g_0)$ represents the difference in NDRGFP1, $(h_1 - h_0)$ represents the difference in NDRGFP2, $(i_1 - i_0)$ represents the difference in IVHX_2, $(j_1 - j_0)$ represents the difference in IVHX_3, $(k_1 - k_0)$ represents the difference in RACE, $(l_1 - l_0)$ represents the difference in TREAT, $(m_1 - m_0)$ represents the difference in SITE, $(n_1 - n_0)$ represents the difference in AGEND1 and $(o_1 - o_0)$ represents the difference in RACESITE.

To save space, not all terms are shown. The expression contains a term for the variance of each main effect (10 terms) as well as an expression for each Covariance (45 terms). All terms involving IVHX_2 simplify to zero, as both subjects have IVHX_2=0.

$$Var\{\ln[OR]\} = 0.3815 = (0.6146745)^2$$

Once the variance is known, the standard error can be obtained by taking its square root and the confidence interval can be formed in the usual manner.

The odds ratio for the subjects described above calculated in Exercise 6 was 8.03. Using the variance calculated above, the 95% confidence interval for this odds ratio is (2.39, 26.93).

```
. lincom (25*_b[age]+5*_b[ndrgfp1]-8.04719*_b[ndrgfp2]+_b[treat]+125*_b[agxfp1]
> )-(30*_b[age]+1.667*_b[ndrgfp1]-.85138*_b[ndrgfp2]+_b[race]+_b[ivhx_3]+_b[sit
> e]+_b[racxssit]+50*_b[agxfp1])

 ( 1)  - 5.0 age + 3.333 ndrgfp1 - 7.19581 ndrgfp2 - ivhx_3 - race + treat - sit
> e + 75.0 agxfp1 - racxssit = 0.0

------------------------------------------------------------------------------
      dfree |    Coef.    Std. Err.      z     P>|z|    [95% Conf. Interval]
------------+-----------------------------------------------------------------
        (1) |   2.082755   .6176745    3.37   0.001    .8721356    3.293375
------------------------------------------------------------------------------


. lincom (25*_b[age]+5*_b[ndrgfp1]-8.04719*_b[ndrgfp2]+_b[treat]+125*_b[agxfp1]
> )-(30*_b[age]+1.667*_b[ndrgfp1]-.85138*_b[ndrgfp2]+_b[race]+_b[ivhx_3]+_b[sit
> e]+_b[racxssit]+50*_b[agxfp1]),or

 ( 1)  - 5.0 age + 3.333 ndrgfp1 - 7.19581 ndrgfp2 - ivhx_3 - race + treat - sit
> e + 75.0 agxfp1 - racxssit = 0.0

------------------------------------------------------------------------------
      dfree | Odds Ratio  Std. Err.      z     P>|z|    [95% Conf. Interval]
------------+-----------------------------------------------------------------
        (1) |   8.026555   4.957798    3.37   0.001    2.392014    26.93362
------------------------------------------------------------------------------
```

8. *We noted in Section 5.5 that an argument could be made for combining the previous and recent levels of IV drug use into one group yielding a dichotomous "never-ever" covariate. Evaluate the model using this covariate in place of history of IV drug use coded at three levels. Is the dichotomous covariate significant? Do the coefficients for the other covariates in the model change? How does the fit of the model compare with the fit of the original model (Table 5.10)?*

Create the dichotomous variable (IVHXD)

```
. gen ivhxd=0 if ivhx==1
 (352 missing values generated)

. replace ivhxd=1 if ivhx>1
 (352 real changes made)

. tab ivhx ivhxd

           |      ivhxd
      ivhx |         0          1 |     Total
-----------+----------------------+----------
         1 |       223          0 |       223
         2 |         0        109 |       109
         3 |         0        243 |       243
-----------+----------------------+----------
     Total |       223        352 |       575
```

Evaluate the model using this covariate in place of IVHX_2 and IVHX_3

```
. logit dfree age  ndrgfp1 ndrgfp2 ivhxd race treat site agend1 racesite

Iteration 0:   Log Likelihood =-326.86446
Iteration 1:   Log Likelihood = -300.0876
Iteration 2:   Log Likelihood =-299.01551
Iteration 3:   Log Likelihood =-299.00872
Iteration 4:   Log Likelihood =-299.00872

Logit Estimates                              Number of obs =      575
                                             chi2(9)       =    55.71
                                             Prob > chi2   =   0.0000
Log Likelihood = -299.00872                  Pseudo R2     =   0.0852


---------------------------------------------------------------------------
   dfree |      Coef.   Std. Err.        z     P>|z|     [95% Conf. Interval]
---------+-----------------------------------------------------------------
     age |    .1162609   .0288081     4.036    0.000      .059798     .1727238
 ndrgfp1 |      1.6705    .407147     4.103    0.000      .872507    2.468494
 ndrgfp2 |    .4350869   .1167357     3.727    0.000     .2062892     .6638847
   ivhxd |   -.6781567   .2346756    -2.890    0.004    -1.138112    -.218201
    race |    .6926185   .2616467     2.647    0.008     .1798003    1.205437
   treat |    .4378541   .2033835     2.153    0.031     .0392297     .8364785
    site |    .5295709   .2484475     2.132    0.033     .0426227    1.016519
  agend1 |   -.0151875   .0060182    -2.524    0.012     -.026983    -.003392
racesite |   -1.443288   .5265047    -2.741    0.006    -2.475218    -.4113575
   _cons |   -6.847807   1.218879    -5.618    0.000    -9.236766    -4.458849
---------------------------------------------------------------------------
```

The dichotomous covariate, IVHXD is significant. When the model coefficients are compared with those from the original model, only very slight changes are seen. The largest change is seen for the variable, SITE, for which the coefficient increases by 2.5%.

Evaluate the fit of the model with the dichotomous covariate

```
. lfit, group (10) table

Logistic model for dfree, goodness-of-fit test
 (Table collapsed on quantiles of estimated probabilities)

_Group     _Prob    _Obs_1    _Exp_1    _Obs_0    _Exp_0    _Total
     1    0.0932         4       4.1        54      53.9        58
     2    0.1262         5       6.2        52      50.8        57
     3    0.1653         7       8.6        51      49.4        58
     4    0.2037        13      10.4        44      46.6        57
     5    0.2343        14      12.7        44      45.3        58
     6    0.2757        11      14.4        46      42.6        57
     7    0.3241        20      17.5        38      40.5        58
     8    0.3737        23      19.8        34      37.2        57
     9    0.4613        23      23.9        35      34.1        58
    10    0.7181        27      29.4        30      27.6        57

          number of observations =       575
                number of groups =        10
        Hosmer-Lemeshow chi2(8) =       4.45
                    Prob > chi2 =      0.8139
```

The fit of the model with the original coding of IVHX was adequate with $p = 0.8199$, there is little difference in the fit of the model with the new coding.

Chapter Six — Solutions

1. *Use the data presented in Breslow and Zhao (1988) to perform the analyses reported in that paper and the analysis reported in Fears and Brown (1986).*

The data presented in Breslow and Zhao are reproduced below:

z_1	Stratum	Sample	Number with $z_2=0$	Number with $z_2=1$	Total Number	Population Frequency
0.60	1	Control	657	45	702	180000.00
		Case	60	11	71	1007.09
	2	Control	191	14	205	102500.00
		Case	107	24	131	2421.44
	3	Control	178	13	191	68214.29
		Case	52	16	68	1611.37
0.14	4	Control	685	40	725	109848.50
		Case	82	26	108	474.52
	5	Control	198	14	212	70666.67
		Case	130	33	163	1000.00
	6	Control	157	16	173	45526.32
		Case	131	30	161	783.45
1.07	7	Control	467	40	507	42996.10
		Case	29	11	40	195.89
	8	Control	131	15	146	31063.83
		Case	63	11	74	431.49
	9	Control	70	7	77	10000.00
		Case	37	7	44	312.72
0.96	10	Control	427	28	455	42523.37
		Case	38	8	46	431.12
	11	Control	138	12	150	27777.78
		Case	104	18	122	1022.63
	12	Control	196	6	202	18035.71
		Case	65	10	75	695.73

To perform the analysis presented in Breslow and Zhao, create the offset term described on page 210 of the text. For example, in stratum 1, the offset is defined as follows:

```
. gen offset=ln((71/1007.09)/(702/180000)) if str==1
(44 missing values generated)
```

To fit a model with an offset in STATA, use the glm command.

```
. glm status z1 z2 [fweight=zfreq], family(binomial) link(logit) offset(offset)

Iteration 1 : deviance = 4876.0521
Iteration 2 : deviance = 4873.7474
Iteration 3 : deviance = 4873.7467
Iteration 4 : deviance = 4873.7467

Residual df   =       4845              No. of obs  =       4848
Pearson X2    =   4757.427              Deviance    =   4873.747
Dispersion    =   .981925              Dispersion  =   1.005933

Bernoulli distribution, logit link, offset offset
-------------------------------------------------------------------------
 status |    Coef.    Std. Err.      z     P>|z|     [95% Conf. Interval]
--------+----------------------------------------------------------------
     z1 |  .5604485   .0925233    6.057   0.000    .3791062    .7417909
     z2 |  1.114423   .1032847   10.790   0.000    .911989    1.316858
  _cons |   -4.8764   .0656913  -74.232   0.000   -5.005153   -4.747648
-------------------------------------------------------------------------
```

The results reported in Fears and Brown differ slightly from those reported in Breslow and Zhao, since Fears and Brown used 60, 14, 107 and 96 as the values of z_1. The analysis using their values is shown below.

```
. glm status z1f z2 [fweight=zfreq], family(binomial) link(logit) offset(offset)

Iteration 1 : deviance = 4876.0521
Iteration 2 : deviance = 4873.7474
Iteration 3 : deviance = 4873.7467
Iteration 4 : deviance = 4873.7467

Residual df   =       4845              No. of obs  =       4848
Pearson X2    =   4757.427              Deviance    =   4873.747
Dispersion    =   .981925              Dispersion  =   1.005933

Bernoulli distribution, logit link, offset offset
-------------------------------------------------------------------------
 status |    Coef.    Std. Err.      z     P>|z|     [95% Conf. Interval]
--------+----------------------------------------------------------------
    z1f |  .0056045   .0009252    6.057   0.000    .0037911    .0074179
     z2 |  1.114423   .1032847   10.790   0.000    .911989    1.316858
  _cons |   -4.8764   .0656913  -74.232   0.000   -5.005153   -4.747648
-------------------------------------------------------------------------
```

2. *Fit the model in Table 6.3 using the suggested fractional polynomials for HSAGEIR and BMPWTLBS. Use this new, non-linear model to provide appropriate odds ratio estimates. Use STATA to import the coefficients from the design-based analysis to compute probabilities of response and assess goodness-of-fit.*

The non-linear model using the suggested transformations from page 219 of the text is shown below.

```
-> gen hsagei_1 = x if e(sample)==1
-> gen hsagei_2 = x^3 if e(sample)==1
where: x = hsageir/10

. gen bmpwt_1=ln(bmpwtlbs)
(54 missing values generated)
```

```
. svylogit hbp  hsagei_1 hsagei_2 bmpwt_1 hssex  race2 race3 bmphtin
(sum of wgt is  1.7695e+008)

Survey logistic regression

pweight:   wtpfhx6                          Number of obs    =      16964
Strata:    sdpstra6                         Number of strata =         49
PSU:       sdppsu6                          Number of PSUs   =         98
                                            Population size  = 1.769e+08
                                            F(  7,     43)   =     152.77
                                            Prob > F         =     0.0000


------------------------------------------------------------------------
     hbp |     Coef.   Std. Err.       t    P>|t|    [95% Conf. Interval]
---------+--------------------------------------------------------------
hsagei_1 |   1.316024   .0852526    15.437   0.000    1.144703   1.487346
hsagei_2 |  -.0051067   .0006545    -7.802   0.000    -.006422  -.0037913
 bmpwt_1 |   1.849155   .1620974    11.408   0.000    1.523408   2.174902
   hssex |    .222497    .078133     2.848   0.006    .0654828   .3795112
   race2 |   .5966679    .073451     8.123   0.000    .4490627   .7442731
   race3 |   .0109612   .3051997     0.036   0.971   -.6023606    .624283
 bmphtin |  -.0677568   .0123544    -5.484   0.000   -.0925839  -.0429297
    _cons |  -12.93879   .8844315   -14.629   0.000   -14.71612  -11.16146
------------------------------------------------------------------------
```

Table. Logistic Regression Analysis of Factors Associated with HBP=1 from the NHANES III Data Set, Non-Linear Model ($n=16,964$).

Variable	Coefficient	Standard Error	Odds Ratio (95% CI)
HSAGEI_1	1.316	0.0853	See table below
HSAGEI_2	-0.005	0.0007	See table below
BMPWT_1	1.849	0.1621	See table below
Male (vs. Female)	0.222	0.0781	1.25 (1.07, 1.46)
Black Race (vs. White)	0.597	0.0735	1.82 (1.57, 2.10)
Other Race (vs. White)	0.011	0.3052	1.01 (0.55, 1.87)
Height (inches)	-0.068	0.0124	0.93 (0.91, 0.96)

*Odds ratio for a 1-inch increase in height.

Table. Odds Ratios for a 10-year increase in AGE at Various Starting Ages.

	AGE=35	AGE=45	AGE=55
Odds Ratio for 10-year Increase in AGE	2.91	2.54	2.15
95% Confidence Interval	(2.60, 3.26)	(2.34, 2.76)	(2.03, 2.27)

$$g(age+10) - g(age) = \left[\beta_1\left(\frac{age+10}{10}\right) + \beta_2\left(\frac{age+10}{10}\right)^3\right] - \left[\beta_1\left(\frac{age}{10}\right) + \beta_2\left(\frac{age}{10}\right)^3\right]$$

$$= \beta_1 + \beta_2\left[\left(\frac{age+10}{10}\right)^3 - \left(\frac{age}{10}\right)^3\right]$$

```
. lincom _b[ HSAG_1]+48.25*_b[ HSAG_2],or

48.25=4.5^3-3.5^3

 ( 1)   HSAG_1 + 48.25 HSAG_2 = 0.0

------------------------------------------------------------------------
      HBP | Odds Ratio   Std. Err.      t    P>|t|    [95% Conf. Interval]
----------+-------------------------------------------------------------
      (1) |   2.914296   .1631844     19.10   0.000    2.604142    3.261389
------------------------------------------------------------------------

. lincom _b[ HSAG_1]+75.25*_b[ HSAG_2],or

75.25=5.5^3-4.5^3

 ( 1)   HSAG_1 + 75.25 HSAG_2 = 0.0

------------------------------------------------------------------------
      HBP | Odds Ratio   Std. Err.      t    P>|t|    [95% Conf. Interval]
----------+-------------------------------------------------------------
      (1) |   2.538945   .1044746     22.64   0.000    2.337441    2.757819
------------------------------------------------------------------------

. lincom _b[ HSAG_1]+108.25*_b[ HSAG_2],or

108.25=6.5^3-5.5^3

 ( 1)   HSAG_1 + 108.25 HSAG_2 = 0.0

------------------------------------------------------------------------
      HBP | Odds Ratio   Std. Err.      t    P>|t|    [95% Conf. Interval]
----------+-------------------------------------------------------------
      (1) |   2.145191   .0598673     27.35   0.000    2.028195    2.268937
------------------------------------------------------------------------
```

Table. Odds Ratios for a 10-pound increase in weight at various starting weights.

	Weight=150	Weight=200	Weight=250
Odds Ratio for 10-pound Increase in Weight	1.13	1.09	1.08
95% Confidence Interval	(1.10, 1.15)	(1.08, 1.11)	(1.06, 1.09)

$$g(wt+10) - g(wt) = \beta \ln(wt+10) - \beta \ln(wt)$$

$$= \beta \ln\left[\left(\frac{wt+10}{wt}\right)\right]$$

```
. lincom 0.0654*_b[ BMPW_1],or

0.0654=ln(160/150)


 ( 1)   .0654 BMPW_1 = 0.0

------------------------------------------------------------------------
      HBP | Odds Ratio  Std. Err.     t    P>|t|    [95% Conf. Interval]
----------+-------------------------------------------------------------
      (1) |   1.128551   .011964   11.41   0.000     1.104763    1.152852
------------------------------------------------------------------------

. lincom 0.04879*_b[ BMPW_1],or

0.04879=ln(210/200)


 ( 1)   .04879 BMPW_1 = 0.0

------------------------------------------------------------------------
      HBP | Odds Ratio  Std. Err.     t    P>|t|    [95% Conf. Interval]
----------+-------------------------------------------------------------
      (1) |   1.094415  .0086554   11.41   0.000     1.077159    1.111948
------------------------------------------------------------------------

0.03922=ln(260/250)

. lincom 0.03922*_b[ BMPW_1],or

 ( 1)   .03922 BMPW_1 = 0.0

------------------------------------------------------------------------
      HBP | Odds Ratio  Std. Err.     t    P>|t|    [95% Conf. Interval]
----------+-------------------------------------------------------------
      (1) |   1.075218  .0068357   11.41   0.000     1.061569    1.089043
------------------------------------------------------------------------
```

Use STATA to import the coefficients from the design-based analysis to compute probabilities of response and assess goodness-of-fit.

```
. mat b = get(_b)

. lfit hbp, beta(b) group (10) outsample

Logistic model for hbp, goodness-of-fit test
(Table collapsed on quantiles of estimated probabilities)

          number of observations =       16964
                 number of groups =          10
         Hosmer-Lemeshow chi2(10) =      108.06
                    Prob > chi2 =       0.0000
```

The goodness-of-fit test using the coefficients from the design-based analysis suggests that the model does not fit. However, due to the large sample size the test has extremely high power. Thus before concluding that the models does not fit we should examine to two by ten table of observed and estimated expected frequencies.

3. *Use the data from the NHANES III survey described in Table 6.1 to find the best model for assessing factors associated with high cholesterol (defined as TCP>230 mg/100mL). Prepare a table of estimated odds ratios and 95 percent confidence intervals for all covariates in the final model. Compare results for the design-based versus the model-based analysis. Assuming the data resulted from a simple random sample, determine whether the continuous covariates in the model are linear in the logit, determine whether there are any significant interactions among the independent variables in the model, assess model calibration and discrimination, and identify poorly fit and influential covariate patterns. Develop your final design-based model taking into consideration all of these aspects of model development.*

The outcome variable HICHOL was created.

```
.  gen HICHOL= (TCP>230)

. replace HICHOL=. if TCP==.
(969 real changes made, 969 to missing)

. tab hichol

     hichol |      Freq.     Percent        Cum.
------------+-----------------------------------
          0 |      11801       73.47       73.47
          1 |       4261       26.53      100.00
------------+-----------------------------------
      Total |      16062      100.00

. sum TCP hichol

    Variable |       Obs        Mean    Std. Dev.       Min        Max
-------------+-------------------------------------------------------
         TCP |     16062    206.0492    44.83707         59        702
      hichol |     16062    .2652845    .4414983          0          1
```

Data were originally available on the weight of each subject in pounds and the height of each subject in inches. Preliminary analysis yielded models with a complex realtionship between high cholesterol and height and weight. As a result height and weight were transformed and combined to create a single variable indicating the body mass index (BMI) of each subject in kg/m^2.

```
. gen htcm=bmphtin*2.54
  (39 missing values generated)

. gen htm=htcm/100
  (39 missing values generated)

. gen wtkg=bmpwtlbs/2.2
(54 missing values generated)

. gen bmi=wtkg/((htm)^2)
(61 missing values generated)
```

The univariate description of all potential covariates did not reveal any variables for which there are illegal values. For each of the categorical variables, at least 1% of cases has a value greater than 0. Binary variables are coded as 0/1. Race (DMARACER) and Smoking Status (SMOKE) are the only non-binary categorical variables. Indicator variables for DMARACER and SMOKE were created.

Bivariate analyses were performed. Each of the categorical variables was cross-classified with HICHOL, producing chi-square statistics. This analysis revealed that each of the categorical variables is significantly associated with the outcome variable. Bivariate analyses of continuous variables were performed using t-tests and Wilcoxon rank-sum tests. Each of the continuous variables is significantly associated with the outcome variable.

The initial multivariable design-based model, shown below, included all potential covariates.

```
. svyset strata sdpstra6

. svyset psu  sdppsu6

. svyset pweight wtpfhx6


---------------------------------------------------------------------------.
svylogit  HICHOL   HSAGEIR  HSSEX  PEPMNK1R  PEPMNK5R  RACE2 RACE3   SMK2 SM
> K3  BMI

Survey logistic regression

pweight:  WTPFHX6                          Number of obs     =       15643
Strata:   SDPSTRA6                         Number of strata  =          49
PSU:      SDPPSU6                          Number of PSUs    =          98
                                           Population size   = 1.664e+08
                                           F(  9,     41)    =       67.40
                                           Prob > F          =      0.0000


-----------------------------------------------------------------------------
     HICHOL |    Coef.     Std. Err.     t     P>|t|   [95% Conf. Interval]
------------+----------------------------------------------------------------
    HSAGEIR |   .032988    .0020954    15.74   0.000    .0287771    .037199
      HSSEX |  -.2876079   .0608624    -4.73   0.000   -.4099155   -.1653003
   PEPMNK1R |   .0044983   .0022366     2.01   0.050    3.67e-06    .0089929
   PEPMNK5R |   .0257226   .0037233     6.91   0.000    .0182402    .0332049
      RACE2 |  -.3036418   .0567075    -5.35   0.000   -.4175998   -.1896837
      RACE3 |  -.1448176   .2083834    -0.69   0.490   -.5635796    .2739445
       SMK2 |   .0293066   .0586465     0.50   0.620   -.0885478    .1471611
       SMK3 |   .108327    .0808078     1.34   0.186   -.0540624    .2707164
        BMI |   .0273513   .0046655     5.86   0.000    .0179756    .036727
      _cons |  -5.746197   .3046832   -18.86   0.000   -6.358481   -5.133913
-----------------------------------------------------------------------------
```

The indicator variables for race and smoking status were deleted one at a time. These variables were neither statistically significant, nor important confounders. The remaining variables, age, sex, systolic blood pressure, diastolic blood pressure and BMI were all statistically significant.

At this point model-based analyses were used to check the scale of the continuous covariates, assess interactions, and evaluate the fit of the model.

```
svylogit  HICHOL   HSAGEIR  HSSEX  PEPMNK1R  PEPMNK5R BMI

Survey logistic regression

pweight:  WTPFHX6                          Number of obs     =       15649
Strata:   SDPSTRA6                         Number of strata  =          49
PSU:      SDPPSU6                          Number of PSUs    =          98
                                           Population size   = 1.664e+08
                                           F(  5,     45)    =      123.40
                                           Prob > F          =      0.0000


-----------------------------------------------------------------------------
     HICHOL |    Coef.     Std. Err.     t     P>|t|   [95% Conf. Interval]
------------+----------------------------------------------------------------
    HSAGEIR |   .0333191   .0020082    16.59   0.000    .0292835    .0373547
      HSSEX |  -.2666884   .0608756    -4.38   0.000   -.3890224   -.1443544
   PEPMNK1R |   .0040831   .0022169     1.84   0.072   -.000372     .0085382
   PEPMNK5R |   .0252709   .0037739     6.70   0.000    .017687     .0328548
        BMI |   .0261643   .0044123     5.93   0.000    .0172974    .0350312
      _cons |  -5.652011   .2911885   -19.41   0.000   -6.237176   -5.066846
-----------------------------------------------------------------------------
```

All of covariates are statistically significant. Four of the five covariates are continuous and the method of fractional polynomials was used to check the scale of these covariates. To minimize confusion, the variables indicating systolic and diastolic blood pressure were renamed SBP and DBP respectively.

The fractional polynomial analysis suggests that the 2, 2 transformation of HSAGEIR significantly improves the model. The 0, 0 transformation of BMI also offers significant improvement. These transformations are shown below. Fractional polynomial analysis showed that the blood pressures indicated that they might be non-linear in the logit. Subsequent analysis using the fracplot option showed minimal departure from linearity. In this case the power to detect even trivial departure from linearity is quite high. One should always check to see that any significant departure from linearity is well supported over the entire range of the covariate and is not being driven by a few observations in the tails. The blood pressure variables will remain in their linear forms.

```
-> gen HSAG_1 = x^2 if _sample
-> gen HSAG_2 = (x^2)*ln(x) if _sample
where: x = hsageir/10

-> gen bmi_1 = ln(x) if _sample
-> gen bmi_2 = ln(x)*ln(x) if _sample
where: x = bmi/10
```

Potential interaction terms were formed using the main effects in the model. These interaction terms were included in the model along with the main effects. The only significant interaction appeared to be between age and BMI. Inclusion of these interaction terms resulted in a poorly calibrated model, for this reason these interaction terms will not be included in future models.

The preliminary final model is shown below.

```
. svylogit  HICHOL  HSAG_1 HSAG_2 HSSEX   SBP DBP BMI_1  BMI_2

Survey logistic regression

pweight:   WTPFHX6                      Number of obs     =       15649
Strata:    SDPSTRA6                     Number of strata  =          49
PSU:       SDPPSU6                      Number of PSUs    =          98
                                        Population size   = 1.664e+08
                                        F(  7,     43)    =      109.72
                                        Prob > F          =      0.0000


------------------------------------------------------------------------------
     HICHOL |     Coef.    Std. Err.      t     P>|t|    [ 95% Conf. Interval]
------------+-----------------------------------------------------------------
     HSAG_1 |   .2823582    .0229662    12.29   0.000    .2362058     .3285105
     HSAG_2 |  -.1167249    .010108    -11.55   0.000   -.1370377    -.0964121
      HSSEX |  -.3612083    .0681523    -5.30   0.000   -.4981655    -.2242511
        SBP |   .0091684    .0024059     3.81   0.000    .0043336     .0140033
        DBP |   .0143633    .0039219     3.66   0.001    .0064819     .0222447
      BMI_1 |   11.07983    1.285666     8.62   0.000    8.496189     13.66348
      BMI_2 |  -4.954638    .6189726    -8.00   0.000   -6.19851     -3.710766
      _cons |  -11.10578    .6432777   -17.26   0.000   -12.39849    -9.813064
------------------------------------------------------------------------------
```

The fit and calibration of the model-based model were examined.

```
. mat b = get(_b)
. lfit HICHOL, beta(b) group (10) outsample table

Logistic model for HICHOL, goodness-of-fit test
(Table collapsed on quantiles of estimated probabilities)

_Group      _Prob     _Obs_1    _Exp_1    _Obs_0    _Exp_0    _Total
     1     0.0806         92      88.8      1473    1476.2      1565
     2     0.1159        157     153.3      1408    1411.7      1565
     3     0.1566        243     212.0      1322    1353.0      1565
     4     0.2062        266     282.8      1299    1282.2      1565
     5     0.2604        356     365.0      1209    1200.0      1565
     6     0.3176        465     452.8      1100    1112.2      1565
     7     0.3744        530     542.5      1035    1022.5      1565
     8     0.4256        589     625.0       976     940.0      1565
     9     0.4873        651     713.0       914     852.0      1565
    10     0.7860        773     853.5       791     710.5      1564

            number of observations =       15649
                  number of groups =          10
         Hosmer-Lemeshow chi2(10) =       37.94
                     Prob > chi2 =         0.0000
```

Although the Hosmer-Lemeshow test is significant, there appears to be reasonable agreement between observed and expected frequencies in the two by ten table. Thus we continue evaluating the fitted model.

```
. mat b=get(_b)

. lroc HICHOL, beta(b)

Logistic model for HICHOL

number of observations =       15649
area under ROC curve    =      0.7035
```

Regression diagnostics based on the model-based, not design-based, fit were used to identify poorly fit and influential covariate patterns. Only the lack-of-fit diagnostic, ΔX^2, was extremely large for four patterns were identified. Each pattern corresponded to a single subject with high cholesterol and a small estimated logistic probability. When these four subjects were deleted none of the coefficients changed by more than 2.1 percent. We do not present the detailed output for this analysis.

Final Model "Model-Based"

```
logit   HICHOL   HSAG_1 HSAG_2 HSSEX    SBP DBP BMI_1   BMI_2

Logit estimates                         Number of obs   =      15649
                                        LR chi2(7)      =    1643.02
                                        Prob > chi2     =     0.0000
Log likelihood = -8201.4963             Pseudo R2       =     0.0910


-------------------------------------------------------------------------
    HICHOL |     Coef.   Std. Err.       z    P>|z|    [95% Conf. Interval]
-----------+-------------------------------------------------------------
    HSAG_1 |   .2859488   .0152265    18.78   0.000    .2561054    .3157922
    HSAG_2 |  -.1177228   .0067777   -17.37   0.000   -.1310068   -.1044388
     HSSEX |  -.3574044   .0400226    -8.93   0.000   -.4358472   -.2789616
       SBP |   .0067985   .0013507     5.03   0.000    .0041512    .0094458
       DBP |    .009452   .0022993     4.11   0.000    .0049454    .0139586
     BMI_1 |      6.379   .7448626     8.56   0.000    4.919097    7.838904
     BMI_2 |  -2.797082   .3570171    -7.83   0.000   -3.496823   -2.097342
     _cons |  -8.073648   .4031391   -20.03   0.000   -8.863786    -7.28351
-------------------------------------------------------------------------
```

Final Model "Design-Based"

```
svylogit    HICHOL   HSAG_1 HSAG_2 HSSEX    SBP DBP BMI_1   BMI_2

Survey logistic regression

pweight:  WTPFHX6                        Number of obs    =       15649
Strata:   SDPSTRA6                       Number of strata =          49
PSU:      SDPPSU6                        Number of PSUs   =          98
                                        Population size   = 1.664e+08
                                        F(   7,      43) =      109.72
                                        Prob > F          =      0.0000


-------------------------------------------------------------------------
    HICHOL |     Coef.   Std. Err.       t    P>|t|    [95% Conf. Interval]
-----------+-------------------------------------------------------------
    HSAG_1 |   .2823582   .0229662    12.29   0.000    .2362058    .3285105
    HSAG_2 |  -.1167249    .010108   -11.55   0.000   -.1370377   -.0964121
     HSSEX |  -.3612083   .0681523    -5.30   0.000   -.4981655   -.2242511
       SBP |   .0091684   .0024059     3.81   0.000    .0043336    .0140033
       DBP |   .0143633   .0039219     3.66   0.001    .0064819    .0222447
     BMI_1 |   11.07983   1.285666     8.62   0.000    8.496189    13.66348
     BMI_2 |  -4.954638   .6189726    -8.00   0.000    -6.19851   -3.710766
     _cons |  -11.10578   .6432777   -17.26   0.000   -12.39849   -9.813064
-------------------------------------------------------------------------
```

Appropriate Odds Ratios and 95% Confidence Intervals for the final model are presented in the table below.

Table. Logistic Regression Analysis of Factors Associated with HICHOL=1 from the NHANES III Data Set, (*n*=15649)

	"Design-Based Analysis"	
Variable	Odds Ratio	95% CI
Age	See table below	
Systolic Blood Pressure	1.10*	1.04, 1.15
Diastolic Blood Pressure	1.15*	1.07, 1.25
Male (vs. Female)	0.69	0.61, 0.80
BMI	See table below	

*Odds ratio for a 10 unit increase in blood pressure

Table. Odds Ratios for a 10-year increase in AGE at Various Starting Ages.

	AGE=35	AGE=45	AGE=55
Design Based			
Odds Ratio for 10-year Increase in AGE	1.62 (1.53, 1.76)	1.43 (1.37, 1.50)	1.19 (1.14, 1.25)

$$g(age+10)-g(age)=\left[\beta_1\left(\frac{age+10}{10}\right)^2+\beta_2\left(\frac{age+10}{10}\right)^2\ln\left(\frac{age+10}{10}\right)\right]-\left[\beta_1\left(\frac{age}{10}\right)^2+\beta_2\left(\frac{age}{10}\right)^2\ln\left(\frac{ag}{10}\right.$$

$$=\beta_1\left[\left(\frac{age+10}{10}\right)^2-\left(\frac{age}{10}\right)^2\right]+\beta_2\left[\left\{\left(\frac{age+10}{10}\right)^2\ln\left(\frac{age+10}{10}\right)\right\}-\left\{\left(\frac{age}{10}\right)^2\ln\left(\frac{age}{10}\right.\right.\right.$$

```
. lincom _b[ HSAG_1]*8+_b[ HSAG_2]*15.11,or

 ( 1)   8.0 HSAG_1 + 15.11 HSAG_2 = 0.0

------------------------------------------------------------------------
      HICHOL | Odds Ratio   Std. Err.       t    P>|t|    [95% Conf. Interval]
-------------+----------------------------------------------------------
         (1) |   1.640748   .0553411    14.68    0.000    1.533221   1.755816

. lincom _b[ HSAG_1]*10+_b[ HSAG_2]*21.11,or

 ( 1)   10.0 HSAG_1 + 21.11 HSAG_2 = 0.0

------------------------------------------------------------------------
      HICHOL | Odds Ratio   Std. Err.       t    P>|t|    [95% Conf. Interval]
-------------+----------------------------------------------------------
         (1) |    1.43264   .0343192    15.01    0.000    1.365307   1.503294
------------------------------------------------------------------------

. lincom _b[ HSAG_1]*12+_b[ HSAG_2]*27.52,or

 ( 1)   12.0 HSAG_1 + 27.52 HSAG_2 = 0.0

------------------------------------------------------------------------
      HICHOL | Odds Ratio   Std. Err.       t    P>|t|    [95% Conf. Interval]
-------------+----------------------------------------------------------
         (1) |   1.192472   .0263863     7.96    0.000    1.140609   1.246694
------------------------------------------------------------------------
```

Table. Odds Ratios for a two unit increase in BMI at various starting BMIs.

	BMI=20	BMI=25	BMI=30
Design Based			
Odds Ratio for 2 unit Increase in BMI	2.74 (2.17, 3.46)	2.28 (1.88, 2.76)	2.10 (1.71, 2.37)

$$g(BMI+2) - g(BMI) = \left[\beta_1 \ln\left(\frac{BMI+2}{10}\right) + \beta_2 \left[\ln\left(\frac{BMI+2}{10}\right)\right]^2 \right] - \left[\beta_1 \ln\left(\frac{BMI}{10}\right) + \beta_2 \left[\ln\left(\frac{BMI}{10}\right)\right]^2 \right]$$

$$= \beta_1 \left[\ln\left(\frac{BMI+2}{BMI}\right)\right] + \beta_2 \left[\ln\left(\frac{BMI+2}{BMI}\right)\right]^2$$

```
. lincom _b[ BMI_1]*0.095+ _b[BMI_2]*0.00903,or

 ( 1)   .095 BMI_1 + .00903 BMI_2 = 0.0

------------------------------------------------------------------------------
      HICHOL | Odds Ratio   Std. Err.      t    P>|t|    [95% Conf. Interval]
-------------+----------------------------------------------------------------
         (1) |   2.739687    .3193849    8.65   0.000    2.167497    3.462927
------------------------------------------------------------------------------

. lincom _b[ BMI_1]*0.077+ _b[BMI_2]*0.00593,or

 ( 1)   .077 BMI_1 + .00593 BMI_2 = 0.0

------------------------------------------------------------------------------
      HICHOL | Odds Ratio   Std. Err.      t    P>|t|    [95% Conf. Interval]
-------------+----------------------------------------------------------------
         (1) |   2.279067    .2172957    8.64   0.000     1.88168    2.760377
------------------------------------------------------------------------------
. lincom _b[ BMI_1]*0.065+ _b[BMI_2]*0.00423,or

 ( 1)   .065 BMI_1 + .00423 BMI_2 = 0.0

------------------------------------------------------------------------------
      HICHOL | Odds Ratio   Std. Err.      t    P>|t|    [95% Conf. Interval]
-------------+----------------------------------------------------------------
         (1) |   2.012205    .1629143    8.64   0.000    1.710062    2.367732
------------------------------------------------------------------------------
```

The models indicate that higher blood pressure is significantly associated with high cholesterol. The association appears to be slightly stronger for diastolic blood pressure than systolic blood pressure. This analysis indicates that men are less likely to have high cholesterol than women. Increasing age and body mass index are also positively associated with high cholesterol. The associations between these covariates and high cholesterol are non-linear. The odds ratios presented above indicate that the effect of increasing age or BMI is strongest among subjects with lower initial values for these covariates.

Chapter Seven – Solutions

1. *Data from the 1-3 matched Benign Breast Disease Study are used in this chapter to illustrate methods for a 1-M matched study. The data are described in Table 7.10. Find the best logistic regression model for a 1-1 matched design using the first of the three controls. (Note: It would have been possible to use any one of the three controls. Designation of the first control was arbitrary.)*

The Benign Breast Disease Study studied 50 women diagnosed with benign breast disease (BBD) as well as three age-matched controls for each case, selected from women who were free of the disease. During an interview, information was collected on a number of potential risk factors for BBD including age at first pregnancy, age at menarche, number of live births, number of stillbirths or miscarriages, weight, age at the last menstrual period, marital status, education and regular medical checkups. This analysis only included information from the case and the first age-matched control.

Information on degree obtained was originally coded at five levels (0=None, 1=High School, 2=Jr. College, 3=College, 4=Masters, 5=Doctoral). A cross-tabulation of this variable with the outcome variable indicated that the numbers of women with college or graduate degrees were small. Before proceeding with any modeling, information on degree was recoded into three levels (0=None, 1=High School, 2=College or Graduate). A similar decision was made for the variable indicating marital status. This variable was originally coded into five categories (1=Married, 2=Divorced, 3=Separated, 4=Widowed, 5=Never Married). For the purposes of this study, it seemed that it might be informative to compare women who had never been married with women who had been married. A new variable (EVMAR) was created that took on only two values (1=Never Married, 0=Ever Married).

Univariate conditional logistic regression models were run (in STATA) for each of the variables in the study. A typical model is shown below:

```
. clogit fndx higd, strata(str)

Iteration 0:  Log Likelihood =-34.421585
Iteration 1:  Log Likelihood =-34.085977
Iteration 2:  Log Likelihood =-34.085717

Conditional (fixed-effects) logistic regression    Number of obs =     100
                                                   chi2(1)       =    1.14
                                                   Prob > chi2   =  0.2850
Log Likelihood = -34.085717                        Pseudo R2     =  0.0165


------------------------------------------------------------------------
    fndx |    Coef.    Std. Err.      z     P>|z|    [95% Conf. Interval]
---------+--------------------------------------------------------------
    higd |  .0798657   .076129    1.049    0.294    -.0693444   .2290758
------------------------------------------------------------------------
```

Likelihood Ratio tests were used to evaluate the significance of each variable. For the Degree variable (DEG), two indicator variables were included in the univariate model and the likelihood ratio test had two degrees-of-freedom. Variables for which the p-value of the likelihood ratio test was less than or equal to 0.10 were selected for initial selection in the multivariable model. The original model contained the variables CHK, AGMN, WT and LIV. This model included only 72 women for whom complete information on all covariates was available. The women who did not have complete information were missing data on LIV, the number of live births. In the original multivariable model, the p-value for the Wald statistic for this variable was 0.627. This variable was dropped from the model and the coefficients for the other variables in the models with and without LIV (still based only on the same 72 women) were compared. The differences in the coefficients for each of the other variables in the model were less than 20% indicating that LIV is not an important confounder. This variable was dropped from the model. The remaining variables CHK, AGMN and WT were considered to be significant with Wald statistic p-values less than 0.10.

At this point, each of the remaining variables was allowed to enter the model (one by one). When the indicator variables for DEG were included in the model, there was nearly a 50% increase in the coefficient for CHK, a 62% increase in the coefficient for AGMN and a 53% change in the coefficient for WT. On the basis of these considerable changes, Degree was considered to be an important confounder. The coefficient and standard error for the degree indicator variables were quite large, suggesting a potential numerical problem. For this reason, categories of the Degree variable were further collapsed. The new variable D01 had two levels (0=No Degree, 1=Some Degree). None of the remaining variables was found to be an important confounder.

AGMN and WT are continuous variables. To assess whether including these variables in their linear form was appropriate, quartile design variables were used and the coefficients were plotted against the midpoints of the quartiles. Also, unconditional fractional polynomial analysis and smoothed plots were used to check the scale of these covariates. There was no indication from any of the scale-checking methods that these variables could not be used in their linear forms.

The final step in the model building process was an evaluation of interactions. Interaction terms were created for each of the variables in the model with AGMT, the matching variable. Also, interaction terms among the main effects were created. None of these interaction terms was found to be significant. The preliminary final model is shown in the table below:

Table 1. The logistic model for the 1-1 matched design from the Benign Breast Disease Data

Variable	Coefficient	Standard Error	Odds Ratio	95% Confidence Interval
Regular Checkups (vs. Yes)	-1.114	0.524	0.33	0.12, 0.92
Age at Menarche (years)	0.554	0.235	1.74*	1.10, 2.76
Weight (pounds)	-0.033	0.016	0.72**	0.53, 0.98
School Degree (vs. No Degree)	1.778	0.959	5.92	0.90, 38.80

*Odds ratio for a one year increase in age at menarche
**Odds ratio for a ten pound increase in weight

To assess model adequacy, the data were "reshaped" to create one record for each matched pair. After creating one record per pair, differences between the case value for a covariate and the control value of the covariate were created for each covariate. Logistic regression models were run using the differences between case and control values as the covariates (this analysis was performed in STATISTIX). The response variable was equal to one for all pairs (difference in FNDX between cases and controls). No constant term was fit. Once it was confirmed that the results of this analysis were the same as the results obtained using the clogit command in STATA, values for leverage, residuals and influence were obtained. Each of these was plotted against the predicted probabilities obtained from the model shown in the table above.

The diagnostic statistics identified pair 12 as having the largest influence on the model. This pair had a leverage value of 0.0276, which was unremarkable. This pair had the highest value for ΔX^2 (43.988) and the largest value for $\Delta\beta$ (0.3127). To assess the sensitivity of the model to this stratum, the model was refit excluding this case-control pair. Results are shown in the table below.

Table 2. Percent change in the coefficients when pair 12 is deleted.

Variable	Coefficient	Delete Pair 12 Coefficient	% Change
Regular Checkups (vs. Yes)	-1.114	-1.617	45.2
Age at Menarche (years)	0.554	0.860	55.2
Weight (pounds)	-0.033	-0.050	51.5
School Degree (vs. No Degree)	1.778	2.637	48.3

The changes in the coefficients that result from excluding pair 12 are substantial. In this pair, the case did not have regular check-ups, the case had an age at menarche that was three years earlier than the control and the case was thirty pounds heavier than the control. Both case and control had a degree. All of these features "go against" the model. A decision about whether to delete this observation should be made based on a decision about the clinical plausibility of the data for this pair.

The model shown in Table 1 suggests that women who have not had regular medical checkups are only 1/3 as likely to be diagnosed as having BBD than women who do have regular medical checkups. Older age at menarche was also associated with an increased likelihood of a BBD diagnosis. For each year that menarche was delayed, the risk of BBD was 1.74 times higher. Weight was found to have a protective effect. Heavier women were less likely to have a diagnosis of BBD. A ten pound increase in weight was associated with a 28% reduction in risk. Education was positively associated with BBD, women with a degree were nearly six times more likely than women without a degree to have the diagnosis. The relatively small number of women in the no degree category contributes to the wide confidence interval for this estimate.

2. *The example in Sections 7.5 through 7.7 used only a few of the variables available in the Benign Breast Disease Study. Repeat the modeling using all the covariates.*

The Benign Breast Disease Study studied 50 women diagnosed with benign breast disease (BBD) as well as three age-matched controls for each case, selected from women who were free of the disease. During an interview, information was collected on a number of potential risk factors for BBD including age at first pregnancy, age at menarche, number of live births, number of stillbirths or miscarriages, weight, age at the last menstrual period, marital status, education and regular medical checkups. This analysis included information from the case and all age-matched controls.

Information on degree obtained was originally coded at five levels (0=None, 1=High School, 2=Jr. College, 3=College, 4=Masters, 5=Doctoral). A cross-tabulation of this variable with the outcome variable indicated that the numbers of women with college or graduate degrees were small. Before proceeding with any modeling, information on degree was recoded into three levels (0=None, 1=High School, 2=College or Graduate). A similar decision was made for the variable indicating marital status. This variable was originally coded into five categories (1=Married, 2=Divorced, 3=Separated, 4=Widowed, 5=Never Married). For the purposes of this study, it seemed that it might be informative to compare women who had never been married with women who had been married. A new variable (EVMAR) was created that took on only two values (1=Never Married, 0=Ever Married).

Univariate conditional logistic regression models were run (in STATA) for each of the variables in the study. A typical model is shown below:

```
. clogit fndx agmn, strata(str)

Iteration 0:   Log Likelihood =-65.783834
Iteration 1:   Log Likelihood =-58.489129
Iteration 2:   Log Likelihood =-58.432973
Iteration 3:   Log Likelihood =-58.432931

Conditional (fixed-effects) logistic regression      Number of obs =    200
                                                     chi2(1)       =  21.76
                                                     Prob > chi2   = 0.0000
Log Likelihood = -58.432931                          Pseudo R2     = 0.1570


-----------------------------------------------------------------------------
    fndx |     Coef.   Std. Err.       z     P>|z|    [ 95% Conf. Interval]
---------+-------------------------------------------------------------------
    agmn |   .4717591   .1109862     4.251   0.000     .2542301     .689288
-----------------------------------------------------------------------------
```

Likelihood Ratio tests were used to evaluate the significance of each variable. For the Degree Variable (DEG), two indicator variables were included in the univariate model and the likelihood ratio test had two degrees-of-freedom. Variables for which the p-value of the likelihood ratio test was less than or equal to 0.10 were selected for initial selection in the multivariable model. The original model contained the variables HIGD, CHK, AGP1, AGMN, WT AGLP and EVMAR. This model included only 148 women for whom complete information on all covariates was available. The women who did not have complete information were missing data on AGP1, the age at first pregnancy. In the original multivariable model, the p-value for the Wald statistic for this variable was 0.321. This variable was dropped from the model and the coefficients for the other variables in the models with and without AGP1 (still based only on the same 148 women) were compared. The differences in the coefficients for each of the other variables in the model were less than 20% indicating that AGP1 is not an important confounder. This variable was dropped from the model. The p-value for the Wald statistic for HIGD was 0.570. When HIGD was excluded from the model, there were no important changes in any of the coefficients for the other variables. The process of deleting nonsignificant variables and evaluating whether the deleted variables were important confounders continued until the model contained only the variables CHK, AGMN, WT and EVMAR.

At this point, each of the remaining variables was allowed to enter the model (one by one). None of the remaining variables was found to be an important confounder.

AGMN and WT are continuous variables. To assess whether including these variables in their linear form was appropriate, quartile design variables were used and the coefficients were plotted against the midpoints of the quartiles. Also, unconditional fractional polynomial analysis and smoothed plots were used to check the scale of these covariates. There was no indication from any of the scale-checking methods that these variables could not be used in their linear forms.

The final step in the model building process was an evaluation of interactions. Interaction terms were created for each of the variables in the model with AGMT, the matching variable. Also, interaction terms among the main effects were created. None of these interaction terms was found to be significant. The preliminary final model is shown in the table below:

Table 3. The logistic model for the 1-3 matched design from the Benign Breast Disease Data

Variable	Coefficient	Standard Error	Odds Ratio	95% Confidence Interval
Regular Checkups (vs. Yes)	-1.161	0.447	0.31	0.13, 0.75
Age at Menarche (years)	0.359	0.128	1.43*	1.11, 1.84
Weight (pounds)	-0.028	0.010	0.76**	0.62, 0.92
Never Married (vs. Ever Married)	1.593	0.736	4.92	1.16, 20.82

*Odds ratio for a one year increase in age at menarche
**Odds ratio for a ten pound increase in weight

The model obtained after considering all covariates in the benign breast disease study is the same as the model considered in Sections 7.5-7.7 of the text. The discussion of model adequacy will not be repeated here.

The model shown in Table 3 suggests that women who did not have regular medical checkups are less than 1/3 as likely to be diagnosed as having BBD than women who do have regular medical checkups. Older age at menarche was also associated with an increased likelihood of a BBD diagnosis. For each year that menarche was delayed, the risk of BBD was 1.43 times higher. Weight was found to have a protective effect. Heavier women were less likely to have a diagnosis of BBD. A ten pound increase in weight was associated with a 24% reduction in risk. Never having been married was associated with nearly 5 times the risk of BBD. The relatively small number of women in the never married category contributes to the wide confidence interval for this estimate.

Chapter Eight – Solutions

1. Data from the mammography experience study are described in Section 8.1. Use a subset of these data and fit a multinomial logistic regression model. For example, you may choose to use only the first 200 subjects. The purpose of this exercise is to obtain practice when there are more than two categories of outcome. Hence, any alternative strategy for identifying a subset of subjects is acceptable.

The Mammography Experience Study studied 412 women to assess factors associated with women's knowledge, attitude and behavior toward mammography. For the purposes of this analysis, a subset of the first 200 subjects was used. This subset was created by selecting subjects whose identification code (OBS) was less than or equal to 200.

The distribution of the outcome variable (ME) in the small subset is shown below:

```
. tab me

        me |      Freq.      Percent        Cum.
-----------+-----------------------------------
         0 |        114        57.00       57.00
         1 |         48        24.00       81.00
         2 |         38        19.00      100.00
-----------+-----------------------------------
     Total |        200       100.00
```

Information on extent of agreement with the statement "You do not need a mammogram unless you develop symptoms" (SYMPT) was originally coded at four levels (1=Strongly Agree, 2=Agree, 3=Disagree, 4=Strongly Disagree). A cross-tabulation of this variable with the outcome variable indicated that the number of women who agreed or strongly agreed with the statement, and who had had mammographies were small. Before proceeding with any modeling, information on SYMPT was recoded into two levels (0=Any Agreement, 1=Any Disagreement). This new variable is SYMPD. This collapsing of categories helps prevent numerical problems in modeling, although the number of women with SYMPD=0 and ME>0 is still relatively small.

A similar process occurred with the variable DETC. This variable, which measured response to the question "How likely is it that a mammogram could find a new case of breast cancer?" was original coded at three levels (1=Not likely, 2=Somewhat likely, 3=Very likely). A cross-tabulation of this variable with the outcome variable indicated that the number of women who did not answer "Very Likely" and who had had a mammogram were quite small. In fact, no women who answered "Not likely" had had a mammogram within the past year. Before proceeding with any modeling, information on DETC was recoded into two levels (0=Not or somewhat likely, 1=Very likely). This new variable is DETCD.

The remaining variables PB, HIST and BSE were originally included in their original forms.

Since there are not very many covariates in this dataset, initial models were built including all five potential covariates. Throughout the modeling process, ME=0 (No mammography experience) is used as the referent category.

```
. mlogit me  sympd pb hist bse detcd


Multinomial regression                              Number of obs =      200
                                                    chi2(10)      =    45.81
                                                    Prob > chi2   =   0.0000
Log Likelihood = -172.78461                         Pseudo R2     =   0.1171


-----------------------------------------------------------------------------
      me |      Coef.    Std. Err.       z      P>|z|     [95% Conf. Interval]
---------+-------------------------------------------------------------------
1        |
   sympd |   2.081733    .6474574     3.215    0.001     .8127394    3.350726
      pb |   -.243593    .1072209    -2.272    0.023    -.4537421   -.0334438
    hist |   1.323921    .5912054     2.239    0.025     .1651796    2.482662
     bse |   1.532375    .8071786     1.898    0.058    -.0496663    3.114416
   detcd |   .4753499     .482882     0.984    0.325    -.4710814    1.421781
   _cons |  -2.677809    1.352569    -1.980    0.048    -5.328795    -.026822
---------+-------------------------------------------------------------------
2        |
   sympd |    1.01054    .4762335     2.122    0.034     .0771396    1.943941
      pb |   -.121587    .1025268    -1.186    0.236    -.3225358     .0793618
    hist |   1.031984    .6265631     1.647    0.100    -.1960572    2.260025
     bse |   .5850873    .5969906     0.980    0.327    -.5849927    1.755167
   detcd |  -.4459084    .4210264    -1.059    0.290    -1.271105     .3792882
   _cons |  -1.204551     1.10753    -1.088    0.277     -3.37527     .9661674
-----------------------------------------------------------------------------
(Outcome me==0 is the comparison group)
```

The Wald statistic was used to identify variables eligible for exclusion from the model. DETCD was the first variable to be excluded from the model. The likelihood ratio test comparing the models with and without this variable indicated that DETCD was not significantly ($p>0.05$) associated with ME. BSE was the second variable to be excluded from the model. The likelihood ratio test comparing the models with and without this variable indicated that BSE was not significantly ($p>0.05$) associated with ME. At this point, the remaining variables in the model, SYMPD, PB and HIST were found to be significantly associated with ME.

DETCD was allowed to reenter the model. Its impact on the coefficients for the other covariates in the model was assessed and was determined to be minor. The impact of BSE on the other coefficients in the model was also found to be minor.

The perceived benefit of mammography (PB) is the only continuous variable in the model. The range of this variable in the subset of data for this analysis is from 5-13. The appropriateness of modeling this variable in its linear form was assessed using two strategies. First, approximate quartiles of PB were generated, corresponding to the values of 5, 6-7, 8-9 and $\geq$10. Three design variables were formed using PB=5 as the reference value. The estimated logistic regression coefficients for the design variables from the two logit functions (from the models also containing SYMPD and HIST) were plotted against the quartile midpoints. This plot is shown below. The plot provides strong evidence that the logits are linear in PB.

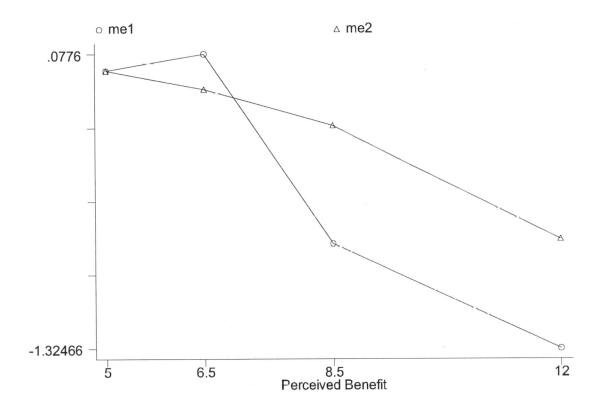

The second strategy used to assess the scale of PB was the use of fractional polynomials when separate binary logistic regression models were fit. The results (shown below) indicate that it is appropriate to use PB in its linear form.

ME=1 vs. ME=0

```
. gen me1v0=me==1

. replace me1v0=. if me==2
(38 real changes made, 38 to missing)
```

```
. fracpoly logit me1v0 pb sympd hist, compare
........
-> gen pb_1 = pb^-2 if _sample
-> gen pb_2 = pb^-2*ln(pb) if _sample

Iteration 0:  Log Likelihood =-98.446335
Iteration 1:  Log Likelihood =-83.263733
Iteration 2:  Log Likelihood =-81.880141
Iteration 3:  Log Likelihood =-81.804703
Iteration 4:  Log Likelihood =-81.804219

Logit Estimates                          Number of obs =     162
                                         chi2(4)       =   33.28
                                         Prob > chi2   =  0.0000
Log Likelihood = -81.804219              Pseudo R2     =  0.1690

------------------------------------------------------------------------
     me |    Coef.   Std. Err.      z    P>|z|    [95% Conf. Interval]
--------+---------------------------------------------------------------
   pb_1 | -344.6959   243.6926   -1.414  0.157   -822.3247    132.9328
   pb_2 |  276.2655   177.4007    1.557  0.119    -71.43352   623.9645
  sympd |  2.095474   .6366971    3.291  0.001      .847571   3.343378
   hist |  1.136088   .600684     1.891  0.059     -.0412308  2.313407
  _cons | -6.375034   2.026688   -3.146  0.002    -10.34727   -2.402799
------------------------------------------------------------------------
Deviance:  163.608. Best powers of pb among 44 models fit: -2 -2.

Fractional Polynomial Model Comparisons:
---------------------------------------------------------------
pb             df      Deviance   Gain    P(term) Powers
---------------------------------------------------------------
Not in model    0       170.150    --       --
Linear          1       164.344   0.000    0.016   1
m = 1           2       163.772   0.573    0.449   3
m = 2           4       163.608   0.736    0.922   -2 -2
---------------------------------------------------------------
```

ME–2 vs. ME=0

```
. gen me2v0=me==2

. replace me2v0=. if me==1
(48 real changes made, 48 to missing)
```

```
. fracpoly logit me2v0 pb sympd hist, compare
........
-> gen pb_1 = pb^3 if _sample
-> gen pb_2 = pb^3*ln(pb) if _sample
```

```
Logit Estimates                          Number of obs  =       152
                                         chi2(4)        =     10.90
                                         Prob > chi2    =    0.0277
Log Likelihood = -80.026257              Pseudo R2      =    0.0637
```

me2v0	Coef.	Std. Err.	z	P>|z|	[95% Conf. Interval]	
pb_1	.0059422	.0112049	0.530	0.596	-.016019	.0279034
pb_2	-.0026785	.0045406	-0.590	0.555	-.0115778	.0062209
sympd	.9212922	.4680888	1.968	0.049	.003855	1.838729
hist	.9754473	.62471	1.561	0.118	-.2489618	2.199856
_cons	-1.919094	.8586212	-2.235	0.025	-3.60196	-.236227

```
Deviance:  160.053. Best powers of pb among 44 models fit: 3 3.
```

Fractional Polynomial Model Comparisons:

pb	df	Deviance	Gain	P(term)	Powers
Not in model	0	162.366	--	--	
Linear	1	160.880	0.000	0.223	1
m = 1	2	160.431	0.449	0.503	3
m = 2	4	160.053	0.827	0.828	3 3

The final step in the model building process was an evaluation of interactions. Interaction terms were created between each of the main effects in the model. There was a significant interaction between PB and HIST in the model comparing ME=2 with ME=0. The preliminary final model is shown below:

```
. mlogit me pb sympd hist pbhist
```

```
Multinomial regression                   Number of obs  =       200
                                         chi2(8)        =     49.81
                                         Prob > chi2    =    0.0000
Log Likelihood = -170.78694              Pseudo R2      =    0.1273
```

me	Coef.	Std. Err.	z	P>|z|	[95% Conf. Interval]	
1						
pb	-.2907056	.1093069	-2.660	0.008	-.5049431	-.076468
sympd	2.140977	.6373188	3.359	0.001	.8918548	3.390098
hist	-2.449325	2.873642	-0.852	0.394	-8.081561	3.182911
pbhist	.56282	.4522784	1.244	0.213	-.3236294	1.449269
_cons	-.599149	.9850266	-0.608	0.543	-2.529766	1.331468
2						
pb	-.2049672	.1074587	-1.907	0.056	-.4155824	.0056479
sympd	.9985998	.4860812	2.054	0.040	.0458981	1.951301
hist	-12.69989	5.76482	-2.203	0.028	-23.99873	-1.401051
pbhist	1.883425	.7580906	2.484	0.013	.3975945	3.369255
_cons	-.3704409	.905499	-0.409	0.682	-2.145186	1.404304

```
(Outcome me==0 is the comparison group)
```

The model shown above indicates that when the PBHIST interaction term is included in the model, there are numerical problems in the ME=2 vs. ME=0 model. Exploration reveals several zero cells. For this reason, the interaction term will not be included in the final model (shown below).

```
. mlogit me pb sympd hist

Iteration 0:   Log Likelihood =-195.69093
Iteration 1:   Log Likelihood =-178.01846
Iteration 2:   Log Likelihood =-176.81518
Iteration 3:   Log Likelihood =-176.75889
Iteration 4:   Log Likelihood = -176.7586
Iteration 5:   Log Likelihood = -176.7586

Multinomial regression                      Number of obs =      200
                                            chi2(6)       =    37.86
                                            Prob > chi2   =   0.0000
Log Likelihood =  -176.7586                 Pseudo R2     =   0.0967

------------------------------------------------------------------------------
      me |      Coef.    Std. Err.       z     P>|z|     [95% Conf. Interval]
---------+--------------------------------------------------------------------
1        |
      pb | -.2833525    .1044565    -2.713    0.007    -.4880835    -.0786214
   sympd |  2.144961     .63775      3.363    0.001     .8949943     3.394928
    hist |  1.278969    .5842426     2.189    0.029     .1338749     2.424064
   _cons | -.6758862    .9554108    -0.707    0.479    -2.548457     1.196685
---------+--------------------------------------------------------------------
2        |
      pb | -.1151296    .0996257    -1.156    0.248    -.3103925     .0801333
   sympd |  .9598075    .4659821     2.060    0.039     .0464993     1.873116
    hist |  1.058617    .6251128     1.693    0.090    -.1665818     2.283815
   _cons | -1.011738    .8710158    -1.162    0.245    -2.718898     .6954216
------------------------------------------------------------------------------
(Outcome me==0 is the comparison group)
```

To assess model adequacy, individual logistic regression models were fit (comparing ME=1 vs. ME=0 and ME=2 vs. ME=0). The individual logistic models are shown below, along with assessments of their overall fit, and any poorly fit or influential covariate patterns.

ME=1 vs. ME=0

```
. logistic me1v0 sympd pb hist

Logit Estimates                                 Number of obs =      162
                                                chi2(3)       =    32.55
                                                Prob > chi2   =   0.0000
Log Likelihood = -82.172083                     Pseudo R2     =   0.1653

------------------------------------------------------------------------
   me10 | Odds Ratio   Std. Err.      z     P>|z|   [95% Conf. Interval]
--------+---------------------------------------------------------------
  sympd |  8.191045    5.221555     3.299   0.001    2.348134     28.573
     pb |  .7837628    .0821988    -2.323   0.020    .6381358   .9626229
   hist |  3.011859    1.804561     1.840   0.066    .9307556   9.746162
------------------------------------------------------------------------
```

The Hosmer-Lemeshow goodness-of-fit test for the ME=1 vs. ME=0 model has a p-value of 0.98, indicating that the fit of the model is adequate.

ME=2 vs. ME=0

```
. logistic me2v0 sympd pb hist

Logit Estimates                                 Number of obs =      152
                                                chi2(3)       =    10.07
                                                Prob > chi2   =   0.0180
Log Likelihood = -80.439926                     Pseudo R2     =   0.0589

------------------------------------------------------------------------
   me20 | Odds Ratio   Std. Err.      z     P>|z|   [95% Conf. Interval]
--------+---------------------------------------------------------------
  sympd |  2.570612    1.20097      2.021   0.043    1.028873   6.422607
     pb |  .8873071    .0881203    -1.204   0.229    .730364    1.077975
   hist |  2.807489    1.747084     1.659   0.097    .8291252   9.506396
------------------------------------------------------------------------
```

The Hosmer-Lemeshow goodness-of-fit test for the ME=2 vs. ME=0 model has a p-value of 0.98, indicating that the fit of the model is adequate.

Regression diagnostics were used to identify covariate patterns that were poorly fit and/or overly influential in both models.   A summary of these patterns is found below:

Data/P#	Logit 1		Logit 2	
	1	2	21	22
SYMPD	0	0	1	1
PB	6	9	5	9
HIST	0	1	1	1
y_j	2	1	0	3
m_j	9	1	2	3
π	0.085	0.119	0.602	0.483
$\Delta\beta$	1.113	0.658	0.871	1.780
ΔX^2	2.967	7.996	3.731	4.474
h	0.273	0.076	0.189	0.285

The data for influential subjects should be evaluated for clinical plausibility.

The final model is presented below:

Table. Final multinomial logistic regression model, ME=0 is the referent category.

Outcome	Variable	Coefficient	Standard Error	Odds Ratio, 95% Confidence Interval
ME=1	Disagreement with Statement "You do not need a mammogram unless you develop symptoms" (vs. agreement)	2.1450	0.6378	8.54 (2.45, 29.81)
	Perceived Benefit of Mammography (Summary Score)	-0.2834	0.1045	1.75* (1.18, 2.63)
	Family History of Breast Cancer (vs. None)	1.2790	0.5842	3.59 (1.14, 11.29)
ME=2	Disagreement with Statement "You do not need a mammogram unless you develop symptoms" (vs. agreement)	0.9598	0.4660	2.61 (1.05, 6.51)
	Perceived Benefit of Mammography (Summary Score)	-0.1151	0.0996	1.26* (0.85, 1.85)
	Family History of Breast Cancer (vs. None)	1.0586	0.6251	2.88 (0.85, 9.81)

*Odds ratio for a two point decrease in the perceived benefit summary score

The analysis reveals that women who disagree with the statement "You do not need a mammogram unless you develop symptoms" are significantly more likely to have mammograms than women who agree with the statement. Women who disagree with the statement were 8.5 times more likely to have had a mammogram within the past year, the confidence interval reveals that this difference may be as small as 2.45 or as large as

29.81. The width of the confidence interval results from a small number of women who agreed with the statement and who had a mammogram within the past year. Women who disagree with the statement were 2.6 times more likely to have had a mammogram in the more distant past, the confidence interval reveals that this difference may be as small as 1.05 or as large as 6.51.

Women who perceive the greatest benefit of mammography are the most likely to have mammograms. A two-point decrease in the perceived benefit summary score (lower values indicate greater perceived benefit) is strongly associated with recent mammography, and with mammography in the more distant past.

Women with a mother or sister with a history of breast cancer are more likely to have had a mammogram than other women. Women with a family history are 3.6 times as likely as women without a family history to have had a mammogram with the past year, the confidence interval suggests that the odds ratio may be as low as 1.14 or as high as 11.29. Women with a family history are 2.9 times as likely as women without a family history to have had a mammogram in the more distant past, the confidence interval suggests that the odds ratio may be as low as 0.85 or as high as 9.81.

In general, stronger associations are seen with more recent (within the past year) mammography than with mammographies in the more distant past.

2. *The data for the low birth weight study are described in Section 1.6.2. These data are used in Section 8.2 to illustrate ordinal logistic regression models via the four category outcome BWT4, (0 if BWT>3500, 1 if 3000<BWT≤3500, 2 if 2500<BWT≤3000, 3 if BWT≤2500). Use the outcome BWT4 and fit the multinomial or baseline logistic regression model.*

The low birth weight study studied 189 women to assess factors associated with low birth weight babies.

The distribution of the outcome variable (BWT4) is shown below:

```
. tab bwt4

     bwt4 |      Freq.      Percent        Cum.
----------+-----------------------------------
        0 |         46        24.34       24.34
        1 |         46        24.34       48.68
        2 |         38        20.11       68.78
        3 |         59        31.22      100.00
----------+-----------------------------------
    Total |        189       100.00
```

Indicator variables for RACE are created. The distribution of history of Premature Labor (PTL) is shown below. Since the number of women with a history of more than one premature labor is quite small, a dichotomous variable PTLD was created. PTLD takes on the value 1 if a woman has a history of one or more premature labors, and the value 0 if there is no history of premature labor. The number of women with more than 1 physician visits during the first trimester is small; therefore, a dichotomous variable FTVD was created. FTVD takes on the value 0 if there were no physician visits during the first trimester, the value 1 if there was one or more physician visits during the first trimester. The remaining variables, AGE, LWT, SMOKE, HT and UI were originally included in their original forms.

Since there are not very many covariates in this dataset, initial models were built including all eight potential covariates. Throughout the modeling process, BWT4=0 (BWT>3500 g) is used as the referent category.

```
. mlogit bwt4  age lwt smoke ht ui race_2 race_3 ptld ftvd, b(0)

Iteration 0:   Log Likelihood =-259.65219
Iteration 1:   Log Likelihood =-225.65519
Iteration 2:   Log Likelihood =-224.05997
Iteration 3:   Log Likelihood =-224.02434
Iteration 4:   Log Likelihood =-224.02428

Multinomial regression                          Number of obs  =      189
                                                chi2(27)       =    71.26
                                                Prob > chi2    =   0.0000
Log Likelihood = -224.02428                     Pseudo R2      =   0.1372

------------------------------------------------------------------------------
    bwt4 |     Coef.   Std. Err.      z     P>|z|    [95% Conf. Interval]
---------+--------------------------------------------------------------------
1        |
     age |   .004013    .045284     0.089   0.929    -.0847419    .0927679
     lwt |   .0019693   .0080138    0.246   0.806    -.0137375    .0176761
   smoke |   1.177091   .537061     2.192   0.028     .1244707    2.229711
      ht |  -1.555195   1.257933   -1.236   0.216    -4.020699    .9103095
      ui |   .6988677   .9031583    0.774   0.439    -1.07129     2.469026
  race_2 |   1.899229   .9426605    2.015   0.044     .0516481    3.746809
  race_3 |   1.484538   .5501499    2.698   0.007     .4062638    2.562812
    ptld |   .4092379   .7952825    0.515   0.607    -1.149487    1.967963
    ftvd |  -.1121459   .468304    -0.239   0.811    -1.030005    .805713
   _cons |  -1.351819   1.441071   -0.938   0.348    -4.176266    1.472628
---------+--------------------------------------------------------------------
2        |
     age |   .0503378   .0463144    1.087   0.277    -.0404368    .1411124
     lwt |  -.0181712   .0094274   -1.927   0.054    -.0366487    .0003062
   smoke |   1.339584   .5647935    2.372   0.018     .2326087    2.446559
      ht |  -.5736614   1.260747   -0.455   0.649    -3.044681    1.897358
      ui |   1.400753   .8666761    1.616   0.106    -.2979013    3.099407
  race_2 |   2.619808   .952561     2.750   0.006     .7528232    4.486794
  race_3 |   .9532494   .609333     1.564   0.118    -.2410214    2.14752
    ptld |  -.5720456   .9060115   -0.631   0.528    -2.347795    1.203704
    ftvd |   .4444725   .5114955    0.869   0.385    -.5580403    1.446985
   _cons |  -.3052647   1.600729   -0.191   0.849    -3.442636    2.832107
---------+--------------------------------------------------------------------
3        |
     age |  -.0146923   .0484616   -0.303   0.762    -.1096752    .0802906
     lwt |  -.0197339   .0089041   -2.216   0.027    -.0371856   -.0022822
   smoke |   1.745506   .5661979    3.083   0.002     .6357785    2.855233
      ht |   1.191557   .8632112    1.380   0.167    -.5003054    2.88342
      ui |   1.52284    .8417363    1.809   0.070    -.1269328    3.172613
  race_2 |   2.883709   .9271573    3.110   0.002     1.066514    4.700904
  race_3 |   1.651054   .5957498    2.771   0.006     .4834057    2.818702
    ptld |   1.191159   .7409627    1.608   0.108    -.2611017    2.643419
    ftvd |  -.0184857   .4859382   -0.038   0.970    -.970907     .9339356
   _cons |   1.113307   1.56636     0.711   0.477    -1.956702    4.183316
------------------------------------------------------------------------------
(Outcome bwt4==0 is the comparison group)
```

The Wald statistic was used to identify variables eligible for exclusion from the model. FTVD was the first variable to be excluded from the model. The likelihood ratio test comparing the models with and without this variable indicated that FTVD was not significantly (p>0.05) associated with BWT4. PTLD was the second variable to be excluded from the model. The likelihood ratio test comparing the models with and without this variable indicated that PTLD was not significantly (p>0.05) associated with BWT4. HT was the third variable to be excluded from the model. The likelihood ratio test comparing the models with and without this variable indicated that HT was not significantly (p>0.05) associated with BWT4. AGE was included in the model as it is generally found to be a clinically important variable, despite the fact that it was not statistically significant. At this point, the remaining variables in the model, LWT, SMOKE, UI and RACE were found to be significantly associated with BWT4.

HT was allowed to reenter the model. Its impact on the coefficients for the other covariates in the model was assessed. Inclusion of this variable resulted in significant changes in the coefficients for AGE and LWT in the BWT4=1 vs. BWT4=0 and BWT4=3 vs. BWT4=0 models. Since it appears that HT is an important confounder, it will be included in the model. PTLD was allowed to reenter the model. Its impact on the coefficients for the other covariates in the model was assessed. Inclusion of this variable resulted in significant changes in the coefficients for AGE and LWT in the BWT4=1 vs. BWT4=0 model and for AGE and UI in the BWT4=3 vs. BWT4=0 model. Since it appears that PTLD is an important confounder, it will be included in the model. The impact of FTVD on the other coefficients in the model was found to be minor. FTVD will not be included in the model.

AGE and LWT are the only continuous variables in the model. The appropriateness of modeling these variables in their linear forms was assessed using two strategies. First, approximate quartiles of AGE and LWT were generated. Three design variables were formed using the first quartile as the reference value. The estimated logistic regression coefficients for the design variables from the three logit functions (from the models also containing the other covariates) were plotted against the quartile midpoints. These plots are shown below. The plots support the decision to use AGE and LWT in their linear forms. Recall that AGE is not significantly associated with BWT, but is included for clinical reasons.

The second strategy used to assess the scale of AGE and LWT was the use of fractional polynomials when separate binary logistic regression models were fit. The results (shown below) indicate that it is appropriate to use both covariates in their linear forms.

BWT4=1 vs. BWT4=0

```
. gen bwt41v0=1 if bwt4==1
  (143 missing values generated)

. replace bwt41v0=0 if bwt4==0
  (46 real changes made)
```

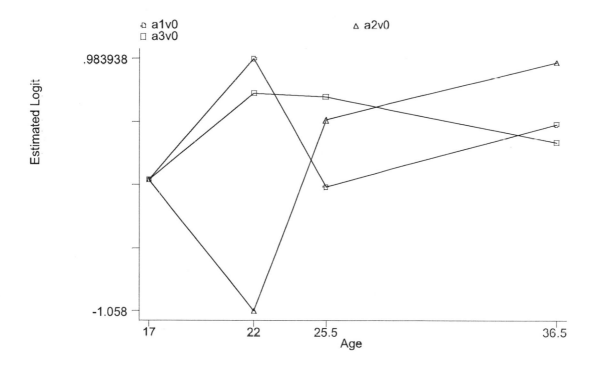

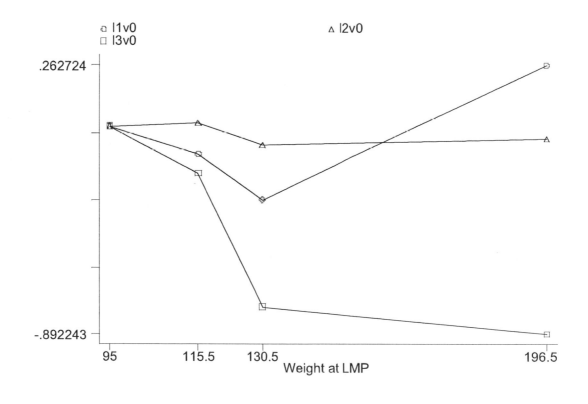

```
. fracpoly logit bwt41v0 age lwt smoke ui race_2 race_3 ht ptld, compare
........
-> gen age_1 = x^-2 if _sample
-> gen age_2 = x^3 if _sample
where: x = age/10

Iteration 0:   Log Likelihood =-63.769541
Iteration 1:   Log Likelihood =-56.100146
Iteration 2:   Log Likelihood =-55.913916
Iteration 3:   Log Likelihood =-55.911975
Iteration 4:   Log Likelihood =-55.911975

Logit Estimates                                  Number of obs =        92
                                                 chi2(9)       =    15.72
                                                 Prob > chi2   =   0.0731
Log Likelihood = -55.911975                      Pseudo R2     =   0.1232


------------------------------------------------------------------------------
bwt41v0 |      Coef.   Std. Err.       z      P>|z|      [95% Conf. Interval]
---------+--------------------------------------------------------------------
  age_1 |  -5.251025   4.660168    -1.127    0.260     -14.38479    3.882735
  age_2 |  -.0335607   .0380304    -0.882    0.378     -.1080989    .0409775
    lwt |   .0009028   .0085274     0.106    0.916     -.0158106    .0176162
  smoke |   .9253193   .5371967     1.722    0.085     -.1275669    1.978205
     ui |   .9992093   .9887064     1.011    0.312     -.9386196    2.937038
 race_2 |   2.101925   1.009164     2.083    0.037      .123999    4.079851
 race_3 |   1.254762   .5316058     2.360    0.018      .2128338    2.29669
     ht |  -1.458122   1.430448    -1.019    0.308     -4.261749    1.345505
   ptld |   .1958869   .8605732     0.228    0.820     -1.490806    1.882579
  _cons |   .5824669   1.905499     0.306    0.760     -3.152243    4.317177
------------------------------------------------------------------------------
Deviance:  111.824. Best powers of age among 44 models fit: -2 3.

Fractional Polynomial Model Comparisons:
------------------------------------------------------------------------------
age              df    Deviance    Gain   P(term) Powers
------------------------------------------------------------------------------
Not in model      0    113.278     --       --
Linear            1    113.218   0.000    0.806   1
m = 1             2    112.809   0.409    0.522   -2
m = 2             4    111.824   1.394    0.611   -2 3
------------------------------------------------------------------------------
```

```
. fracpoly logit bwt41v0 lwt age smoke ui race_2 race_3 ht ptld, compare
........
-> gen lwt_1 = x^-2 if _sample
-> gen lwt_2 = x^-2*ln(x) if _sample
where: x = lwt/100

Iteration 0:   Log Likelihood =-63.769541
Iteration 1:   Log Likelihood =-56.449092
Iteration 2:   Log Likelihood =-56.326442
Iteration 3:   Log Likelihood =-56.325835
Iteration 4:   Log Likelihood =-56.325835

Logit Estimates                              Number of obs =       92
                                             chi2(9)       =    14.89
                                             Prob > chi2   =   0.0941
Log Likelihood = -56.325835                  Pseudo R2     =   0.1167

------------------------------------------------------------------------------
bwt41v0 |      Coef.   Std. Err.       z     P>|z|      [95% Conf. Interval]
---------+--------------------------------------------------------------------
  lwt_1 |  -1.451679    2.476177    -0.586   0.558     -6.304897    3.401538
  lwt_2 |  -6.112443    8.371744    -0.730   0.465    -22.52076    10.29587
    age |   .0090436    .0432841     0.209   0.834     -.0757916    .0938788
  smoke |   .7621947    .5495034     1.387   0.165     -.3148122   1.839202
     ui |   .6696624    1.060416     0.632   0.528     -1.408715    2.74804
 race_2 |   1.884446    .9765806     1.930   0.054     -.0296165   3.790509
 race_3 |   1.238072    .5339886     2.319   0.020      .1914739   2.284671
     ht |  -1.531723    1.444804    -1.060   0.289     -4.363488   1.300041
   ptld |   .3939642    .8442869     0.467   0.641     -1.260808   2.040736
  _cons |   .6242071    2.947198     0.212   0.832     -5.152195    6.40061
------------------------------------------------------------------------------
Deviance:  112.652. Best powers of lwt among 44 models fit: -2 -2.

Fractional Polynomial Model Comparisons:
------------------------------------------------------------------------------
lwt                df      Deviance     Gain   P(term) Powers
------------------------------------------------------------------------------
Not in model        0      113.237      --       --
Linear              1      113.218     0.000    0.892   1
m = 1               2      113.181     0.037    0.847   3
m = 2               4      112.652     0.566    0.768   -2 -2
------------------------------------------------------------------------------
```

BWT4=2 vs. BWT4=0

```
. gen bwt42v0=1 if bwt4==2
 (151 missing values generated)

. replace bwt42v0=0 if bwt4==0
 (46 real changes made)
```

```
. fracpoly logit bwt42v0 age lwt smoke ui race_2 race_3 ht ptld, compare
........
-> gen age_1 = x^3 if _sample
-> gen age_2 = x^3*ln(x) if _sample
where: x = age/10

Iteration 0:  Log Likelihood =-57.842833
Iteration 1:  Log Likelihood =-44.854589
Iteration 2:  Log Likelihood =-44.123378
Iteration 3:  Log Likelihood =-44.102491
Iteration 4:  Log Likelihood =-44.102458

Logit Estimates                              Number of obs =       84
                                             chi2(9)       =    27.48
                                             Prob > chi2   =   0.0012
Log Likelihood = -44.102458                  Pseudo R2     =   0.2375


------------------------------------------------------------------------------
 bwt42v0 |     Coef.    Std. Err.       z      P>|z|     [95% Conf. Interval]
---------+--------------------------------------------------------------------
   age_1 |  .3873785    .2035229      1.903    0.057     -.011519     .786276
   age_2 | -.2454361    .1419162     -1.729    0.084     -.5235868    .0327146
     lwt | -.0263949    .0120765     -2.186    0.029     -.0500643   -.0027255
   smoke |  2.036494    .7344784      2.773    0.006      .5969432    3.476045
      ui |  1.290452    .916008       1.409    0.159     -.5048907    3.085795
  race_2 |  3.749299    1.239155      3.026    0.002      1.320601    6.177998
  race_3 |  1.35314     .7006921      1.931    0.053     -.0201916    2.726471
      ht | -.3423494    1.428164     -0.240    0.811      -3.1415     2.456801
    ptld | -1.944972    1.251382     -1.554    0.120     -4.397634    .5076913
   _cons | -.4061672    1.759414     -0.231    0.817     -3.854555    3.042221
------------------------------------------------------------------------------
Deviance:   88.205. Best powers of age among 44 models fit: 3 3.

Fractional Polynomial Model Comparisons:
-----------------------------------------------------------------
age            df      Deviance     Gain    P(term)  Powers
-----------------------------------------------------------------
Not in model    0      94.002        --        --
Linear          1      90.173       0.000     0.050   1
m = 1           2      88.846       1.327     0.249   -2
m = 2           4      88.205       1.968     0.726   3 3
-----------------------------------------------------------------
```

```
    fracpoly logit bwt42v0 lwt age smoke ui race_2 race_3 ht ptld, compare
........
-> gen lwt_1 = x^-2 if _sample
-> gen lwt_2 = x^-2*ln(x) if _sample
where: x = lwt/100

Iteration 0:   Log Likelihood =-57.842833
Iteration 1:   Log Likelihood =-44.148753
Iteration 2:   Log Likelihood =-43.422164
Iteration 3:   Log Likelihood =-43.404761
Iteration 4:   Log Likelihood =-43.404746

Logit Estimates                            Number of obs =       84
                                           chi2(9)       =    28.88
                                           Prob > chi2   =   0.0007
Log Likelihood = -43.404746                Pseudo R2     =   0.2496

------------------------------------------------------------------------------
 bwt42v0 |     Coef.   Std. Err.       z     P>|z|     [95% Conf. Interval]
---------+--------------------------------------------------------------------
   lwt_1 |  -.8935296   3.661473    -0.244   0.807    -8.069884    6.282825
   lwt_2 |   -15.0221    12.5377    -1.198   0.231    -39.59555    9.551349
     age |   .1189159   .0541833     2.195   0.028     .0127185    .2251132
   smoke |    2.11742   .7430361     2.850   0.004     .6610962    3.573744
      ui |   1.368225   .9575509     1.429   0.153    -.5085402     3.24499
  race_2 |   3.359107   1.120296     2.998   0.003     1.163366    5.554847
  race_3 |   1.530414   .7350731     2.082   0.037     .0896972    2.971131
      ht |  -2.215602   1.662412    -1.333   0.183     -5.47387    1.042667
    ptld |  -2.866681   1.478817    -1.938   0.053    -5.765109    .0317473
   _cons |  -2.027779   4.058057    -0.500   0.617    -9.981424    5.925867
------------------------------------------------------------------------------
Deviance:   86.809. Best powers of lwt among 44 models fit: -2 -2.

Fractional Polynomial Model Comparisons:
------------------------------------------------------------------------
lwt                df      Deviance     Gain    P(term)  Powers
------------------------------------------------------------------------
Not in model        0       94.763       --       --
Linear              1       90.173     0.000     0.032    1
m = 1               2       88.273     1.899     0.168    -2
m = 2               4       86.809     3.363     0.481    -2 -2
------------------------------------------------------------------------
```

BWT4=3 vs. BWT4=0

```
. gen bwt43v0=1 if bwt4==3
 (130 missing values generated)

. replace bwt43v0=0 if bwt4==0
 (46 real changes made)
```

```
. fracpoly logit bwt43v0 age lwt smoke ui race_2 race_3 ht ptld, compare
........
-> gen age_1 = x^3 if _sample
-> gen age_2 = x^3*ln(x) if _sample
where: x = age/10

Iteration 0:  Log Likelihood =-71.973623
Iteration 1:  Log Likelihood = -49.45137
Iteration 2:  Log Likelihood = -46.96007
Iteration 3:  Log Likelihood =-46.620515
Iteration 4:  Log Likelihood =-46.565606
Iteration 5:  Log Likelihood =-46.564723
Iteration 6:  Log Likelihood =-46.564723

Logit Estimates                              Number of obs =      105
                                             chi2(9)       =    50.82
                                             Prob > chi2   =   0.0000
Log Likelihood = -46.564723                  Pseudo R2     =   0.3530


------------------------------------------------------------------------
 bwt43v0 |     Coef.    Std. Err.      z      P>|z|    [95% Conf. Interval]
---------+--------------------------------------------------------------
   age_1 |  .3209259    .3935508     0.815    0.415    -.4504194   1.092271
   age_2 |  -.260608    .3125891    -0.834    0.404    -.8732715    .3520554
     lwt |  -.0334555   .0123349    -2.712    0.007    -.0576316   -.0092795
   smoke |  1.949139    .6972545     2.795    0.005     .5825454   3.315733
      ui |  1.796933    .9696025     1.853    0.064    -.1034536   3.697319
  race_2 |  3.341902    1.160426     2.880    0.004     1.067509   5.616294
  race_3 |  1.512927    .6871783     2.202    0.028     .1660823   2.859772
      ht |  1.613834    1.053273     1.532    0.125    -.4505426   3.678211
    ptld |  1.069736    .8824471     1.212    0.225     -.659828   2.799301
   _cons |  1.443005    2.033183     0.710    0.478    -2.541961   5.427971
------------------------------------------------------------------------
Deviance:   93.129. Best powers of age among 44 models fit: 3 3.

Fractional Polynomial Model Comparisons:
-----------------------------------------------------------------
age             df      Deviance    Gain    P(term) Powers
-----------------------------------------------------------------
Not in model    0       94.248      --       --
Linear          1       94.210     0.000    0.845   1
m = 1           2       94.040     0.171    0.679   3
m = 2           4       93.129     1.081    0.634   3 3
-----------------------------------------------------------------
```

```
. fracpoly logit bwt43v0 lwt age smoke ui race_2 race_3 ht ptld, compare
........
-> gen lwt_1 = x^-2 if _sample
-> gen lwt_2 = x^3 if _sample
where: x = lwt/100

Iteration 0:   Log Likelihood =-71.973623
Iteration 1:   Log Likelihood =-49.366681
Iteration 2:   Log Likelihood =-46.931272
Iteration 3:   Log Likelihood =-46.724337
Iteration 4:   Log Likelihood =-46.721903
Iteration 5:   Log Likelihood =-46.721902

Logit Estimates                           Number of obs  =     105
                                          chi2(9)        =   50.50
                                          Prob > chi2    =  0.0000
Log Likelihood = -46.721902               Pseudo R2      =  0.3508


------------------------------------------------------------------------------
 bwt43v0 |    Coef.    Std. Err.      z     P>|z|     [95% Conf. Interval]
---------+--------------------------------------------------------------------
   lwt_1 |  1.466212   1.792778    0.818    0.413    -2.047568    4.979993
   lwt_2 | -.3211773   .2380498   -1.349    0.177    -.7877464    .1453919
     age | -.0099816   .0547956   -0.182    0.855    -.117379     .0974159
   smoke |  1.887759   .6943419    2.719    0.007     .5268736    3.248644
      ui |  1.707151   .9362727    1.823    0.068    -.1279103    3.542211
  race_2 |  3.392774   1.243571    2.728    0.006     .9554196    5.830128
  race_3 |  1.557442   .6983659    2.230    0.026     .18867      2.926214
      ht |  1.802043   1.135751    1.587    0.113    - 4239884    4.020074
    ptld |  1.159765   .8433936    1.375    0.169    -.4932557    2.812787
   _cons | -1.794053   2.123278   -0.845    0.398    -5.955602    2.367496
------------------------------------------------------------------------------
Deviance:   93.444. Best powers of lwt among 44 models fit: -2 3.

Fractional Polynomial Model Comparisons:
-----------------------------------------------------------
lwt              df     Deviance    Gain    P(term) Powers
-----------------------------------------------------------
Not in model      0     103.029      --       --
Linear            1      94.210     0.000    0.003   1
m = 1             2      94.019     0.191    0.662   2
m = 2             4      93.444     0.767    0.750   -2 3
-----------------------------------------------------------
```

The fractional polynomial analysis supports the decision to include the variables AGE and LWT in their linear forms.

The final step in the model building process was an evaluation of interactions. Interaction terms were created between each of the main effects in the model. There was a significant interaction between AGE and SMOKE in the models comparing BWT4=1 with BWT4=0 and BWT4=3 vs. BWT4=0. However, when this interaction term is included in the model, there are numerical problems in all three models. Exploration reveals several zero cells when AGE and SMOKE are cross-classified within levels of the outcome variable. For this reason, the interaction term will not be included in the final model (shown below).

```
. mlogit bwt4 age lwt smoke ui race_2 race_3 ht ptld, b(0)

Iteration 0:   Log Likelihood =-259.65219
Iteration 1:   Log Likelihood =-226.47536
Iteration 2:   Log Likelihood =-224.81706
Iteration 3:   Log Likelihood =-224.77889
Iteration 4:   Log Likelihood =-224.77882

Multinomial regression                          Number of obs =     189
                                                chi2(24)      =   69.75
                                                Prob > chi2   =  0.0000
Log Likelihood = -224.77882                     Pseudo R2     =  0.1343


--------------------------------------------------------------------------
    bwt4 |    Coef.    Std. Err.      z      P>|z|    [95% Conf. Interval]
---------+----------------------------------------------------------------
group1   |
     age |  .0018409    .0443682    0.041    0.967   -.0851191     .0888008
     lwt |   .001748    .0079785    0.219    0.827   -.0138897     .0173856
   smoke |  1.190816    .5313271    2.241    0.025    .1494343     2.232198
      ui |  .7103551    .9024704    0.787    0.431   -1.058454     2.479165
  race_2 |  1.895927      .94424    2.008    0.045    .0452509     3.746604
  race_3 |  1.497788    .5388446    2.780    0.005    .4416717     2.553904
      ht | -1.511052    1.251547   -1.207    0.227   -3.964039     .9419359
    ptld |  .3972094    .7938864    0.500    0.617   -1.158779     1.953198
   _cons | -1.332812    1.430962   -0.931    0.352   -4.137445     1.471822
---------+----------------------------------------------------------------
group2   |
     age |  .0586839     .045168    1.299    0.194   -.0298437     .1472115
     lwt | -.0183144    .0093648   -1.956    0.051   -.0366692     .0000403
   smoke |  1.253162    .5569992    2.250    0.024    .1614642     2.344861
      ui |  1.372962    .8673766    1.583    0.113   -.3270651     3.072989
  race_2 |   2.58736    .9516639    2.719    0.007    .7221335     4.452587
  race_3 |  .8274922    .5905004    1.401    0.161   -.3298672     1.984852
      ht | -.6507674    1.270206   -0.512    0.608   -3.140326     1.838791
    ptld | -.5313691    .9038116   -0.588    0.557   -2.302807     1.240069
   _cons | -.1635117    1.588746   -0.103    0.918   -3.277397     2.950373
---------+----------------------------------------------------------------
group3   |
     age |  -.015123    .0475814   -0.318    0.751   -.1083809     .0781349
     lwt | -.0200696    .0088733   -2.262    0.024   -.0374609    -.0026783
   smoke |  1.758574    .5539606    3.175    0.002    .6728311     2.844317
      ui |  1.545206    .8416328    1.836    0.066   -.1043639     3.194776
  race_2 |  2.905598    .9278029    3.132    0.002    1.087138     4.724058
  race_3 |  1.653974    .5786637    2.858    0.004    .5198143     2.788134
      ht |  1.211998    .8624495    1.405    0.160   -.4783717     2.902368
    ptld |   1.18291    .7394437    1.600    0.110   -.2663734     2.632193
   _cons |  1.146399    1.552759    0.738    0.460   -1.896952      4.18975
--------------------------------------------------------------------------
(Outcome bwt4==group0 is the comparison group)
```

To assess model adequacy, individual logistic regression models were fit (comparing BWT4 = 1 vs. BWT4 = 0, BWT4 = 2 vs. BWT4 = 0 and BWT4 = 3 vs. BWT4 = 0). The individual logistic models are shown below, along with assessments of their overall fit, and any poorly fit or influential covariate patterns.

BWT4=1 vs. BWT4=0

```
. logit bwt41v0 age lwt smoke ui race_2 race_3 ht ptld, nolog

Logit estimates                                Number of obs   =         92
                                               LR chi2(8)      =      14.32
                                               Prob > chi2     =     0.0738
Log likelihood = -56.608997                    Pseudo R2       =     0.1123

------------------------------------------------------------------------------
     bwt41v0 |      Coef.   Std. Err.       z     P>|z|    [95% Conf. Interval]
-------------+----------------------------------------------------------------
         age |   .0106227   .0432994     0.25    0.806    -.0742426    .0954881
         lwt |   .0011448   .0084211     0.14    0.892    -.0153602    .0176498
       smoke |   .8814013   .5278354     1.67    0.095    -.1531371     1.91594
          ui |   .9609709    .979884     0.98    0.327    -.9595664    2.881508
      race_2 |   1.997919   .9840106     2.03    0.042     .0692936    3.926544
      race_3 |   1.289268   .5275431     2.44    0.015     .2553026    2.323234
          ht |  -1.310427   1.402669    -0.93    0.350    -4.059607    1.438752
        ptld |   .2955267   .8341829     0.35    0.723    -1.339442    1.930495
       _cons |  -1.316628   1.518048    -0.87    0.386    -4.291947    1.658692
------------------------------------------------------------------------------
```

The Hosmer-Lemeshow goodness-of-fit test for the BWT4 = 1 vs. BWT4 = 0 model has a *p*-value of 0.68, indicating that the fit of the model is adequate. The area under the ROC curve is 0.73, indicating acceptable discrimination.

BWT4=2 vs. BWT4=0

```
. logit bwt42v0 age lwt smoke ui race_2 race_3 ht ptld, nolog

Logit estimates                                Number of obs   =         84
                                               LR chi2(8)      =      25.51
                                               Prob > chi2     =     0.0013
Log likelihood = -45.086452                    Pseudo R2       =     0.2205

------------------------------------------------------------------------------
     bwt42v0 |      Coef.   Std. Err.       z     P>|z|    [95% Conf. Interval]
-------------+----------------------------------------------------------------
         age |   .0999991   .0532687     1.88    0.060    -.0044057    .2044039
         lwt |  -.0239984   .0117029    -2.05    0.040    -.0469357   -.0010612
       smoke |   1.961967   .7167451     2.74    0.006     .5571723    3.366761
          ui |   1.381576   .9135261     1.51    0.130    -.4089022    3.172054
      race_2 |    3.49071   1.190262     2.93    0.003      1.15784     5.82358
      race_3 |   1.345164   .7000541     1.92    0.055    -.0269165    2.717245
          ht |  -.3268598   1.396205    -0.23    0.815    -3.063372    2.409652
        ptld |  -1.703252   1.215286    -1.40    0.161    -4.085169    .6786652
       _cons |  -.8452651   1.890586    -0.45    0.655    -4.550745    2.860215
------------------------------------------------------------------------------
```

The Hosmer-Lemeshow goodness-of-fit test for the BWT4=2 vs. BWT4=0 model has a p-value of 0.09, indicating that the fit of the model is adequate. The area under the ROC curve is 0.80, indicating excellent discrimination.

BWT4=3 vs. BWT4=0

```
. logit bwt43v0 age lwt smoke ui race_2 race_3 ht ptld, nolog

Logit estimates                               Number of obs   =        105
                                              LR chi2(8)      =      49.74
                                              Prob > chi2     =     0.0000
Log likelihood = -47.105176                   Pseudo R2       =     0.3455

---------------------------------------------------------------------------
     bwt43v0 |      Coef.   Std. Err.       z    P>|z|     [95% Conf. Interval]
-------------+-------------------------------------------------------------
         age |  -.0106056    .0545253    -0.19   0.846    -.1174731    .096262
         lwt |  -.0322236    .0119131    -2.70   0.007    -.0555729   -.0088743
       smoke |   1.920389    .6916533     2.78   0.005     .5647739   3.276005
          ui |    1.71167    .9352134     1.83   0.067     -.121315   3.544654
      race_2 |   3.275393   1.146871      2.86   0.004     1.027567    5.52322
      race_3 |   1.580895    .6930621     2.28   0.023      .222518   2.939271
          ht |   1.640972   1.045246      1.57   0.116    -.4076734   3.689617
        ptld |   1.183788    .8421676     1.41   0.160    -.4668299   2.834406
       _cons |   2.561266   2.024957      1.26   0.206    -1.407577   6.530109
---------------------------------------------------------------------------
```

The Hosmer-Lemeshow goodness-of-fit test for the BWT4 = 3 vs. BWT4 = 0 model has a *p*-value of 0.12, indicating that the fit of the model is adequate. The area under the ROC curve is 0.87, indicating excellent discrimination.

Regression diagnostics were used to identify covariate patterns that were poorly fit and/or overly influential in both models. A summary of these patterns is found below:

Data/P#	Logit 1			
	1	2	3	4
AGE	18	22	25	25
LWT	90	120	241	95
SMOKE	1	0	0	1
UI	1	0	0	1
RACE	1	1	2	1
HT	0	1	1	0
PTLD	0	0	0	1
BWT4	1	1	0	0
y_j	2	1	0	0
m_j	2	1	1	1
π	0.694	0.095	0.478	0.768
$\Delta\beta$	1.114	2.693	1.623	0.964
ΔX^2	1.524	11.73	1.761	4.085
h	0.422	0.187	0.480	0.191

Data/P#	Logit 2							
	1	2	3	4	5	6	7	8
AGE	22	24	25	45	25	25	28	36
LWT	95	90	95	123	241	155	167	202
SMOKE	0	1	1	0	0	0	0	0
UI	0	0	1	0	0	0	0	0
RACE	3	1	1	1	2	1	1	1
HT	1	0	0	0	1	0	0	0
PTLD	0	1	1	0	0	0	0	0
BWT4	2	2	0	0	0	2	2	2
y_j	1	1	0	0	0	1	1	1
m_j	1	1	1	1	1	1	1	1
π	0.523	0.414	0.734	0.669	0.276	0.113	0.114	0.110
$\Delta\beta$	2.958	1.140	1.546	0.942	0.674	0.290	0.336	0.862
ΔX^2	2.159	2.160	3.864	2.719	0.732	8.166	8.114	8.895
h	0.578	0.345	0.286	0.257	0.480	0.034	0.040	0.088

Data/P#	Logit 3		
	1	2	3
AGE	25	17	24
LWT	95	120	138
SMOKE	1	1	0
UI	1	0	0
RACE	1	3	1
HT	0	0	0
PTLD	1	0	0
BWT4	0	0	3
y_j	0	0	1
m_j	1	1	1
π	0.983	0.882	0.105
$\Delta\beta$	1.572	0.554	0.262
ΔX^2	58.97	8.022	8.755
h	0.026	0.065	0.029

The data for influential subjects should be evaluated for clinical plausibility.

The final model is presented below:

Table. Final Model.

Outcome	Variable	Coefficient	Standard Error	Odds Ratio, 95% Confidence Interval
BWT4=1	Smoking During Pregnancy (vs. No smoking)	1.191	0.5313	3.29 (1.16, 9.32)
	Presence of Uterine Irritability (vs. None)	0.710	0.9025	2.03 (0.35, 11.93)
	History of Premature Labor (vs. No History)	0.397	0.794	1.49 (0.31, 7.05)
	History of Hypertension (vs. No History)	-1.151	1.2515	0.22 (0.02, 2.56)
	Black Race (vs. White)	1.896	0.9442	6.66 (1.05, 42.38)
	Other Race (vs. White)	1.498	0.5388	4.47 (1.56, 12.86)
	Weight of Mother at Last Menstrual Period (pounds)	0.002	0.0080	1.02* (0.87,1.19)
	Age (years)	0.002	0.0443	1.02** (0.43, 2.43)
BWT4=2	Smoking During Pregnancy (vs. No smoking)	1.253	0.5570	3.50 (1.18, 10.43)
	Presence of Uterine	1.373	0.8674	3.95 (0.72, 1.38)

	Irritability (vs. None)			
	History of Premature Labor (vs. No History)	-0.531	0.9038	0.59 (0.10, 3.46)
	History of Hypertension (vs. No History)	-0.651	1.2702	0.52 (0.04, 6.29)
	Black Race (vs. White)	2.587	0.9517	13.29 (2.06, 85.85)
	Other Race (vs. White)	0.827	0.5905	2.29 (0.72, 7.28)
	Weight of Mother at Last Menstrual Period (pounds)	-0.018	0.0094	0.83* (0.69, 1.00)
	Age (years)	0.059	0.0452	1.80** (0.74, 4.36)
BWT4=3	Smoking During Pregnancy (vs. No smoking)	1.759	0.5540	5.80 (1.96, 17.19)
	Presence of Uterine Irritability (vs. None)	1.545	0.8416	4.69 (0.90, 24.40)
	History of Premature Labor (vs. No History)	1.183	0.7394	3.26 (0.77, 13.90)
	History of Hypertension (vs. No History)	1.212	0.8624	3.36 (0.62, 18.22)
	Black Race	2.906	0.9278	18.28

(vs. White)			(2.97, 112.62)
Other Race (vs. White)	1.654	0.5787	5.23 (1.68, 16.25)
Weight of Mother at Last Menstrual Period (pounds)	-0.020	0.0089	0.82* (0.69, 0.97)
Age (years)	-0.015	0.0476	0.86** (0.34, 2.18)

*Odds ratio for a 10-pound increase in weight.
**Odds ratio for a 10-year increase in age.

The analysis reveals that mother's smoking during pregnancy is significantly associated with increased risk of having a lower birthweight baby. Women who smoke during pregnancy are nearly 6 times more likely than non-smokers to have a baby in the lowest quartile of birthweight and are more than 3 times as likely than non-smokers to have babies in the middle quartiles of birthweight.

Black women are more than 18 times more likely than white women to have a baby in the lowest quartile of birthweight. This dramatic increase in risk is also seen for babies in the middle quartiles of birthweight. The same relation is seen for women of other races, although the magnitude of the effect is not as large

Weight of the mother at the last menstrual period is inversely associated with lower birthweight. Lowest birthweight babies are most likely to have low weight mothers.

Presence of uterine irritability, history of hypertension and history of preterm labor are all associated with increased risk of having a baby in the lowest birthweight quartile. Although only uterine irritability appears to be associated with increased risk of a middle quartile baby.

3. Using the final models identified in problems 1 and 2 compare the estimates of the coefficients obtained from fitting the multinomial logistic regression model to those obtained from fitting the adjacent-category, continuation-ratio and proportional odds ordinal logistic regression models. For the mammography experience data recode the outcome variable, ME to 0= Never, 1= Over one year ago and 2= Within one year in order that its codes increase with frequency of use.

Mammography Experience Data

```
. gen menew=0 if me==0
(86 missing values generated)

. replace menew=1 if me==2
(38 real changes made)

. replace menew=2 if me==1
(48 real changes made)
```

The final multinomial logistic regression model with MENEW:

```
. mlogit menew sympd pb hist

Iteration 0:  Log Likelihood =-195.69093
Iteration 1:  Log Likelihood =-178.01846
Iteration 2:  Log Likelihood =-176.81518
Iteration 3:  Log Likelihood =-176.75889
Iteration 4:  Log Likelihood = -176.7586
Iteration 5:  Log Likelihood = -176.7586

Multinomial regression                     Number of obs =      200
                                           chi2(6)       =    37.86
                                           Prob > chi2   = 0.0000
Log Likelihood =  -176.7586                Pseudo R2     = 0.0967

------------------------------------------------------------------------------
   menew |      Coef.   Std. Err.       z     P>|z|     [95% Conf. Interval]
---------+--------------------------------------------------------------------
1        |
   sympd |   .9598075   .4659821      2.060   0.039      .0464993    1.873116
      pb |  -.1151296   .0996257     -1.156   0.248     -.3103925    .0801333
    hist |   1.058617   .6251128      1.693   0.090     -.1665818    2.283815
   _cons |  -1.011738   .8710158     -1.162   0.245     -2.718898    .6954216
---------+--------------------------------------------------------------------
2        |
   sympd |   2.144961    .63775       3.363   0.001      .8949943    3.394928
      pb |  -.2833525   .1044565     -2.713   0.007     -.4880835   -.0786214
    hist |   1.278969   .5842426      2.189   0.029      .1338749    2.424064
   _cons |  -.6758862   .9554108     -0.707   0.479     -2.548457    1.196685
------------------------------------------------------------------------------
(Outcome menew==0 is the comparison group)
```

The adjacent-category model:

```
. constraint define 1 [recent]sympd = 2*[past]sympd

. constraint define 2 [recent]pb = 2*[past]pb

. constraint define 3 [recent]hist = 2*[past]hist
```

```
. mlogit menew sympd pb hist, c(1-3)

 ( 1) - 2.0 [past]sympd + [recent]sympd = 0.0
 ( 2) - 2.0 [past]pb + [recent]pb = 0.0
 ( 3) - 2.0 [past]hist + [recent]hist = 0.0

Iteration 0:  Log Likelihood =-195.69093
Iteration 1:  Log Likelihood =-178.35875
Iteration 2:  Log Likelihood =-177.12955
Iteration 3:  Log Likelihood =-177.09855
Iteration 4:  Log Likelihood =-177.09851

Multinomial regression                        Number of obs =      200
                                              chi2(3)       =    37.18
                                              Prob > chi2   =   0.0000
Log Likelihood = -177.09851                   Pseudo R2     =   0.0950


------------------------------------------------------------------------------
   menew |      Coef.   Std. Err.       z     P>|z|     [95% Conf. Interval]
---------+--------------------------------------------------------------------
past     |
   sympd |   1.042629    .2765446     3.770   0.000     .5006118    1.584647
      pb |  -.1389888    .0505985    -2.747   0.006    -.2381601   -.0398175
    hist |   .6223726    .2807453     2.217   0.027     .072122    1.172623
   _cons |   -.849472    .4781393    -1.777   0.076    -1.786608    .0876639
---------+--------------------------------------------------------------------
recent   |
   sympd |   2.085258    .5530892     3.770   0.000     1.001224    3.169293
      pb |  -.2779776    .1011971    -2.747   0.006    -.4763203    -.079635
    hist |   1.244745    .5614905     2.217   0.027     .1442441    2.345246
   _cons |  -.6621785    .8919139    -0.742   0.458    -2.410298    1.085941
------------------------------------------------------------------------------
(Outcome menew==never is the comparison group)
```

The unconstrained continuation-ratio model:

```
. gen ystar1=1 if menew==1
 (162 missing values generated)

. replace ystar1=0 if menew==0
 (114 real changes made)
```

```
. logit ystar1 sympd pb hist

Iteration 0:   Log Likelihood =-85.474942
Iteration 1:   Log Likelihood =-80.627209
Iteration 2:   Log Likelihood =-80.440317
Iteration 3:   Log Likelihood =-80.439926

Logit Estimates                              Number of obs =      152
                                             chi2(3)       =    10.07
                                             Prob > chi2   =   0.0180
Log Likelihood = -80.439926                  Pseudo R2     =   0.0589

------------------------------------------------------------------------
  ystar1 |    Coef.   Std. Err.      z     P>|z|    [95% Conf. Interval]
---------+--------------------------------------------------------------
   sympd |   .944144   .4671923    2.021   0.043     .028464    1.859824
      pb |  -.1195642  .0993121   -1.204   0.229    -.3142122   .0750839
    hist |   1.03229   .6222943    1.659   0.097    -.1873841   2.251965
   _cons |  -.9650231  .8613064   -1.120   0.263    -2.653153   .7231065
------------------------------------------------------------------------
```

```
. gen ystar2=1 if menew==2
(152 missing values generated)

. replace ystar2=0 if menew==0|menew==1
(152 real changes made)
```

```
. logit ystar2 sympd pb hist

Iteration 0:   Log Likelihood =-110.21599
Iteration 1:   Log Likelihood =-97.720331
Iteration 2:   Log Likelihood =-96.447978
Iteration 3:   Log Likelihood =-96.384167
Iteration 4:   Log Likelihood =-96.383777
Iteration 5:   Log Likelihood =-96.383777

Logit Estimates                              Number of obs =      200
                                             chi2(3)       =    27.66
                                             Prob > chi2   =   0.0000
Log Likelihood = -96.383777                  Pseudo R2     =   0.1255

------------------------------------------------------------------------
  ystar2 |    Coef.   Std. Err.      z     P>|z|    [95% Conf. Interval]
---------+--------------------------------------------------------------
   sympd |  1.912767   .6281857    3.045   0.002     .6815457   3.143988
      pb |  -.2493016  .0997558   -2.499   0.012    -.4448193  -.0537838
    hist |   .8447372  .4910177    1.720   0.085    -.1176397   1.807114
   _cons |  -1.042572  .9265264   -1.125   0.260    -2.85853    .7733867
------------------------------------------------------------------------
```

The proportional odds ordinal model:

```
. ologit menew sympd pb hist

Iteration 0:   Log Likelihood =-195.69093
Iteration 1:   Log Likelihood =-177.56456
Iteration 2:   Log Likelihood =-177.01927
Iteration 3:   Log Likelihood =-177.01529
Iteration 4:   Log Likelihood =-177.01529

Ordered Logit Estimates                    Number of obs  =      200
                                           chi2(3)        =    37.35
                                           Prob > chi2    = 0.0000
Log Likelihood = -177.01529                Pseudo R2      = 0.0954


------------------------------------------------------------------------
   menew |    Coef.    Std. Err.      z      P>|z|    [95% Conf. Interval]
---------+--------------------------------------------------------------
   sympd |  1.52688    .3922496    3.893    0.000     .7580851   2.295675
      pb |  -.22885    .0783067   -2.922    0.003    -.3823283  -.0753718
    hist | 1.055244    .4374256    2.412    0.016     .1979059   1.912583
---------+--------------------------------------------------------------
   _cut1 | -.150868    .6740691           (Ancillary parameters)
   _cut2 | .8569536    .6758372
------------------------------------------------------------------------
```

Table of Estimated Coefficients Under Four Different Multinomial Logistic Models for the Sample of 200 Subjects from the Mamography Experience Study

Variable/Logit	Baseline		Model Continuation Ratio		Adjacent Category	Proportional Odds
	1	2	1	2	1	1
sympd	0.960	2.145	0.944	1.913	1.043	1.527
pb	-0.115	-0.283	-0.120	-0.249	-0.139	-0.229
hist	1.059	1.279	1.032	0.844	0.622	1.055

All four models show an effect due to sympd for ME = k versus ME = $k - 1$ with a log-odds of about 1.0. The log odds for ME = 2 vs ME <2 is about 2.0. The log-odds from the proportional odds model is close to the average log-odds from the other models. The estimated coefficients for pb follow similar patterns as the log-odds for sympd. The patterns do not seem to hold up for the estimated log-odds for family history. In particular the log-odds for the second continuation ratio logit are not larger but smaller than the first continuation logit. The log-odds for the adjacent category model is smaller than the first baseline logit; but twice this value is close the log-odds for the second baseline logit. The log-odds under the proportional odds model is nearly identical to the log-odds from the first baseline logit. The coding for ME used in this problem is not strongly ordinal as it conveys only a general ranking of use of mammography in the past.

In order to decide which model fits best one needs to perform various likelihood ratio tests and assess overall fit.

Low Birthweight Data

The final multinomial logistic regression model:

```
. mlogit bwt4 age lwt smoke ui race_2 race_3 ht ptld, b(0)

Iteration 0:  Log Likelihood =-259.65219

Iteration 1:  Log Likelihood =-226.47536
Iteration 2:  Log Likelihood =-224.81706
Iteration 3:  Log Likelihood =-224.77889
Iteration 4:  Log Likelihood =-224.77882

Multinomial regression                    Number of obs =      189
                                          chi2(24)      =    69.75
                                          Prob > chi2   =   0.0000
Log Likelihood = -224.77882               Pseudo R2     =   0.1343

------------------------------------------------------------------------------
     bwt4 |     Coef.    Std. Err.      z     P>|z|    [95% Conf. Interval]
----------+-------------------------------------------------------------------
group1    |
      age |  .0018409    .0443682     0.041   0.967    -.0851191     .0888008
      lwt |   .001748    .0079785     0.219   0.827    -.0138897     .0173050
    smoke |  1.190816    .5313271     2.241   0.025     .1494343     2.232198
       ui |  .7103551    .9024704     0.787   0.431    -1.058454     2.479165
   race_2 |  1.895927     .94424      2.008   0.045     .0452509     3.746604
   race_3 |  1.497788    .5388446     2.780   0.005     .4416717     2.553904
       ht | -1.511052    1.251547    -1.207   0.227    -3.964039     .9419359
     ptld |  .3972094    .7938864     0.500   0.617    -1.158779     1.953198
    _cons | -1.332812    1.430962    -0.931   0.352    -4.137445     1.471822
----------+-------------------------------------------------------------------
group2    |
      age |  .0586839     .045168     1.299   0.194    -.0298437     .1472115
      lwt | -.0183144    .0093648    -1.956   0.051    -.0366692     .0000403
    smoke |  1.253162    .5569992     2.250   0.024     .1614642     2.344861
       ui |  1.372962    .8673766     1.583   0.113    -.3270651     3.072989
   race_2 |   2.58736    .9516639     2.719   0.007     .7221335     4.452587
   race_3 |  .8274922    .5905004     1.401   0.161    -.3298672     1.984852
       ht | -.6507674    1.270206    -0.512   0.608    -3.140326     1.838791
     ptld | -.5313691    .9038116    -0.588   0.557    -2.302807     1.240069
    _cons | -.1635117    1.588746    -0.103   0.918    -3.277397     2.950373
----------+-------------------------------------------------------------------
group3    |
      age |  -.015123    .0475814    -0.318   0.751    -.1083809     .0781349
      lwt | -.0200696    .0088733    -2.262   0.024    -.0374609    -.0026783
    smoke |  1.758574    .5539606     3.175   0.002     .6728311     2.844317
       ui |  1.545206    .8416328     1.836   0.066    -.1043639     3.194776
   race_2 |  2.905598    .9278029     3.132   0.002     1.087138     4.724058
   race_3 |  1.653974    .5786637     2.858   0.004     .5198143     2.788134
       ht |  1.211998    .8624495     1.405   0.160    -.4783717     2.902368
     ptld |   1.18291    .7394437     1.600   0.110    -.2663734     2.632193
    _cons |  1.146399    1.552759     0.738   0.460    -1.896952     4.18975
------------------------------------------------------------------------------
(Outcome bwt4==group0 is the comparison group)
```

The adjacent-category model:

```
. constraint define 1 [group2]age= 2*[group1]age, mod
. constraint define 2 [group3]age=3*[group1]age, mod
. constraint define 3 [group2]lwt = 2*[group1]lwt
. constraint define 4 [group3]lwt=3*[group1]lwt
. constraint define 5 [group2]smoke= 2*[group1]smoke
. constraint define 6 [group3]smoke=3*[group1]smoke
. constraint define 7 [group2]ui= 2*[group1]ui
. constraint define 8 [group3]ui=3*[group1]ui
. constraint define 9 [group2]race_2= 2*[group1]race_2
. constraint define 10   [group3]race_2=3*[group1]race_2
. constraint define 11 [group2]race_3= 2*[group1]race_3
. constraint define 12   [group3]race_3=3*[group1]race_3
. constraint define 13 [group2]ht= 2*[group1]ht
. constraint define 14   [group3]ht=3*[group1]ht
. constraint define 15 [group2]ptld= 2*[group1]ptld
. constraint define 16   [group3]ptld=3*[group1]ptld
```

```
. mlogit bwt4   age lwt smoke ui race_2 race_3 ht ptld, b(0) c(1-16)
 ( 1) - 2.0 [group1]age + [group2]age = 0.0
 ( 2) - 3.0 [group1]age + [group3]age = 0.0
 ( 3) - 2.0 [group1]lwt + [group2]lwt = 0.0
 ( 4) - 3.0 [group1]lwt + [group3]lwt = 0.0
 ( 5) - 2.0 [group1]smoke + [group2]smoke = 0.0
 ( 6) - 3.0 [group1]smoke + [group3]smoke = 0.0
 ( 7) - 2.0 [group1]ui + [group2]ui = 0.0
 ( 8) - 3.0 [group1]ui + [group3]ui = 0.0
 ( 9) - 2.0 [group1]race_2 + [group2]race_2 = 0.0
 (10) - 3.0 [group1]race_2 + [group3]race_2 = 0.0
 (11) - 2.0 [group1]race_3 + [group2]race_3 = 0.0
 (12) - 3.0 [group1]race_3 + [group3]race_3 = 0.0
 (13) - 2.0 [group1]ht + [group2]ht = 0.0
 (14) - 3.0 [group1]ht + [group3]ht = 0.0
 (15) - 2.0 [group1]ptld + [group2]ptld = 0.0
 (16) - 3.0 [group1]ptld + [group3]ptld = 0.0
Iteration 0:   Log Likelihood =-259.65219
Iteration 1:   Log Likelihood =-236.62392
Iteration 2:   Log Likelihood =-235.25198
Iteration 3:   Log Likelihood =-235.23659
Iteration 4:   Log Likelihood =-235.23659
```

Multinomial regression

```
                                          Number of obs =      189
                                          chi2(8)       =    48.83
                                          Prob > chi2   =   0.0000
Log Likelihood =  235.23659               Pseudo R2     =   0.0940
```

bwt4	Coef.	Std. Err.	z	P>|z|	[95% Conf. Interval]	
group1						
age	.0014818	.014514	0.102	0.919	-.0269651	.0299288
lwt	-.0079286	.0027384	-2.895	0.004	-.0132958	.0025615
smoke	.5035104	.1694469	2.971	0.003	.1714006	.8356201
ui	.4820201	.2193021	2.198	0.028	.0521958	.9118443
race_2	.8117034	.244753	3.316	0.001	.3319964	1.29141
race_3	.4333353	.1789393	2.422	0.015	.0826208	.7840499
ht	.6162307	.3166299	1.946	0.052	-.0043526	1.236814
ptld	.3931999	.2190579	1.795	0.073	-.0361456	.8225455
_cons	.5962363	.5327751	1.119	0.263	-.4479836	1.640456
group2						
age	.0029637	.029028	0.102	0.919	-.0539301	.0598575
lwt	-.0158573	.0054768	-2.895	0.004	-.0265916	-.005123
smoke	1.007021	.3388937	2.971	0.003	.3428012	1.67124
ui	.9640401	.4386043	2.198	0.028	.1043916	1.823689
race_2	1.623407	.489506	3.316	0.001	.6639927	2.582821
race_3	.8666707	.3578786	2.422	0.015	.1652416	1.5681
ht	1.232461	.6332599	1.946	0.052	-.0087052	2.473628
ptld	.7863998	.4381158	1.795	0.073	-.0722913	1.645091
_cons	.7895624	.9870636	0.800	0.424	-1.145047	2.724172
group3						
age	.0044455	.043542	0.102	0.919	-.0808952	.0897863
lwt	-.0237859	.0082152	-2.895	0.004	-.0398873	-.0076845
smoke	1.510531	.5083406	2.971	0.003	.5142018	2.50686
ui	1.44606	.6579064	2.198	0.028	.1565873	2.735533
race_2	2.43511	.734259	3.316	0.001	.9959891	3.874231
race_3	1.300006	.5368179	2.422	0.015	.2478623	2.35215
ht	1.848692	.9498898	1.946	0.052	-.0130578	3.710442
ptld	1.1796	.6571736	1.795	0.073	-.1084369	2.467636
_cons	1.390502	1.442963	0.964	0.335	-1.437652	4.218657

```
(Outcome bwt4==group0 is the comparison group)
```

The continuation-ratio model:

```
. gen ystar1=1 if bwt4==1

. replace ystar1=0 if bwt4==0
(46 real changes made)

. gen ystar2=1 if bwt4==2
(151 missing values generated)

. replace ystar2=0 if bwt4==0|bwt4==1
(92 real changes made)

. gen ystar3=1 if bwt4==3
(130 missing values generated)

. replace ystar3=0 if
bwt4==0|bwt4==1|bwt4==2
(130 real changes made)
```

```
. logit ystar1 age lwt smoke ui race_2 race_3 ht ptld

Iteration 0:  Log Likelihood =-63.769541
Iteration 1:  Log Likelihood =-56.714094
Iteration 2:  Log Likelihood =-56.609341
Iteration 3:  Log Likelihood =-56.608997
Iteration 4:  Log Likelihood =-56.608997

Logit Estimates                      Number of obs =       92
                                     chi2(8)       =    14.32
                                     Prob > chi2   =   0.0738
Log Likelihood = -56.608997          Pseudo R2     =   0.1123

------------------------------------------------------------------------------
  ystar1 |     Coef.    Std. Err.       z     P>|z|    [95% Conf. Interval]
---------+--------------------------------------------------------------------
     age |   .0106227   .0432994     0.245    0.806    -.0742426     .0954881
     lwt |   .0011448   .0084211     0.136    0.892    -.0153602     .0176498
   smoke |   .8814013   .5278354     1.670    0.095    -.1531371      1.91594
      ui |   .9609709    .979884     0.981    0.327    -.9595664     2.881508
  race_2 |   1.997919   .9840106     2.030    0.042     .0692936     3.926544
  race_3 |   1.289268   .5275431     2.444    0.015     .2553026     2.323234
      ht |  -1.310427   1.402669    -0.934    0.350    -4.059607     1.438752
    ptld |   .2955267   .8341829     0.354    0.723    -1.339442     1.930495
   _cons |  -1.316628   1.518048    -0.867    0.386    -4.291947     1.658692
------------------------------------------------------------------------------
```

```
. logit ystar2 age lwt smoke ui race_2 race_3 ht ptld

Iteration 0:   Log Likelihood =-78.546655
Iteration 1:   Log Likelihood =-70.790801
Iteration 2:   Log Likelihood =-70.583493
Iteration 3:   Log Likelihood =-70.582664
Iteration 4:   Log Likelihood =-70.582664

Logit Estimates                              Number of obs =      130
                                             chi2(8)       =    15.93
                                             Prob > chi2   =   0.0434
Log Likelihood = -70.582664                  Pseudo R2     =   0.1014


------------------------------------------------------------------------------
  ystar2 |     Coef.    Std. Err.       z     P>|z|    [95% Conf. Interval]
---------+--------------------------------------------------------------------
     age |   .0645542    .0400949     1.610   0.107   -.0140304    .1431388
     lwt |  -.0186793    .0082789    -2.256   0.024   -.0349056    -.002453
   smoke |   .6139345    .4630876     1.326   0.185   -.2937006     1.52157
      ui |   1.048794    .6383085     1.643   0.100   -.2022677    2.299855
  race_2 |   1.663753     .681423     2.442   0.015    .3281883    2.999317
  race_3 |   .1029789    .4968121     0.207   0.836    -.870755    1.076713
      ht |   .2326534    1.207305     0.193   0.847   -2.133621    2.598928
    ptld |  -1.032069    .8040052    -1.284   0.199    -2.60789    .5437525
   _cons |  -.4992147    1.385555    -0.360   0.719   -3.214852    2.216423
------------------------------------------------------------------------------
```

```
. logit ystar3 age lwt smoke ui race_2 race_3 ht ptld

Iteration 0:   Log Likelihood =  -117.336
Iteration 1:   Log Likelihood =-99.218282
Iteration 2:   Log Likelihood =-98.428105
Iteration 3:   Log Likelihood =-98.416855
Iteration 4:   Log Likelihood =-98.416852

Logit Estimates                              Number of obs =      189
                                             chi2(8)       =    37.84
                                             Prob > chi2   =   0.0000
Log Likelihood = -98.416852                  Pseudo R2     =   0.1612


------------------------------------------------------------------------------
  ystar3 |     Coef.    Std. Err.       z     P>|z|    [95% Conf. Interval]
---------+--------------------------------------------------------------------
     age |  -.0377496    .0378109    -0.998   0.318   -.1118577    .0363584
     lwt |  -.0149103    .0070405    -2.118   0.034   -.0287093   -.0011112
   smoke |   .8464023    .4080745     2.074   0.038    .0465909    1.646214
      ui |   .7111278    .4631199     1.536   0.125   -.1965706    1.618826
  race_2 |   1.212742    .5324883     2.277   0.023    .1690841      2.2564
  race_3 |   .8041194    .4484447     1.793   0.073    -.074816    1.683055
      ht |   1.838687    .7032521     2.615   0.009    .4603384    3.217036
    ptld |   1.221751    .4630153     2.639   0.008    .3142574    2.129244
   _cons |   .6369097    1.230312     0.518   0.605   -1.774458    3.048277
------------------------------------------------------------------------------
```

The proportional odds ordinal model:

```
.  ologit bwt4 age lwt smoke ui race_2 race_3 ht ptld

Iteration 0:   Log Likelihood =-259.65219
Iteration 1:   Log Likelihood =-235.91069
Iteration 2:   Log Likelihood =-235.65121
Iteration 3:   Log Likelihood =-235.65042

Ordered Logit Estimates                        Number of obs =       189
                                               chi2(8)       =     48.00
                                               Prob > chi2   =    0.0000
Log Likelihood = -235.65042                    Pseudo R2     =    0.0924

------------------------------------------------------------------------------
    bwt4 |      Coef.   Std. Err.       z     P>|z|    [95% Conf. Interval]
---------+--------------------------------------------------------------------
     age | -.0006257    .0274693    -0.023   0.982    -.0544645     .0532131
     lwt | -.0128958    .0048733    -2.646   0.008    -.0224473    -.0033442
   smoke |  .9877202    .3149779     3.136   0.002     .3703748     1.605066
      ui |  .9129658    .4044862     2.257   0.024     .1201874     1.705744
  race_2 |  1.470897    .4346912     3.384   0.001     .6189183     2.322876
  race_3 |   .869222    .3344913     2.599   0.009     .2136312     1.524813
      ht |     1.194    .6122466     1.950   0.051    -.0059809     2.393982
    ptld |  .8219579    .4173644     1.969   0.049     .0039387     1.639977
---------+--------------------------------------------------------------------
   _cut1 | -1.803489    .8913835            (Ancillary parameters)
   _cut2 | -.5160991    .8816949
   _cut3 |  .4952642      .87984
------------------------------------------------------------------------------
```

To complete this problem create a table similar to the one used to summarize the fitted multinomial models to the mammography experience data.

4. *The following exercise is designed to enhance the idea expressed in Figure 8.2 and Figure 8.3 that one way to obtain the proportional odds model is via categorization of a continuous variable.*

    (a) Form the scatterplot of BWT vs. LWT

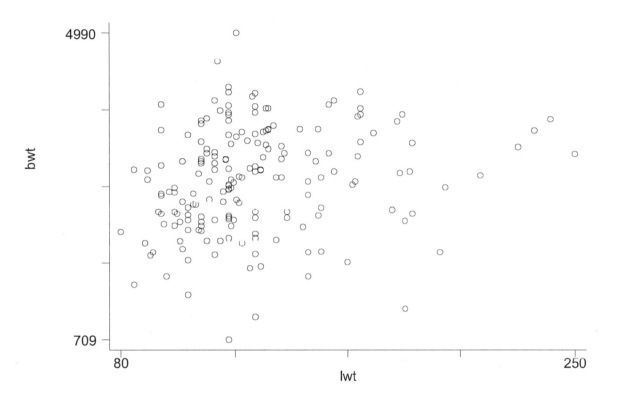

    (b) Fit the linear regression of BWT on LWT and add the estimated regression line to the scatterplot in 4(a). Let λ_0 denote the estimate of the intercept, λ_1, the estimate of the slope and s the root mean squared error from the linear regression.

```
. fit bwt lwt

  Source |       SS       df       MS                  Number of obs =      189
---------+------------------------------              F(  1,    187) =     6.69
   Model | 3448881.30       1   3448881.30            Prob > F       =   0.0105
Residual | 96468171.3     187   515872.574            R-squared      =   0.0345
---------+------------------------------              Adj R-squared  =   0.0294
   Total | 99917052.6     188   531473.684            Root MSE       =   718.24


---------------------------------------------------------------------------------
     bwt |      Coef.   Std. Err.       t     P>|t|     [95% Conf. Interval]
---------+-----------------------------------------------------------------------
     lwt |   4.429264   1.713025     2.586    0.010     1.049927      7.8086
   _cons |   2369.672   228.4306    10.374    0.000      1919.04    2820.304
---------------------------------------------------------------------------------
```

$$\lambda_0 = 2369.672 \qquad \lambda_1 = 4.429264 \qquad s = 718.24$$

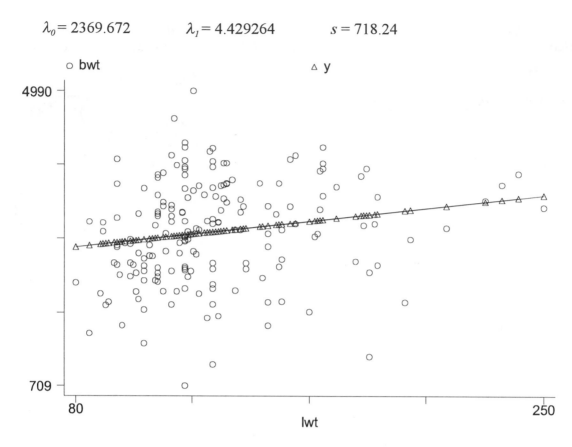

(c) It follows from results for the logistic distribution that the relationship between the root mean squared error in the normal errors linear regression and the scale parameter for logistic errors linear regression is approximately σ=s*(sqrt(3))/π. Use the results from the linear regression in 4(b) and obtain σ.

$$\hat{\sigma} = s\sqrt{3} \,/\, \pi$$

$$\hat{\sigma} = 395.99$$

(d) Use the results from 4(b) and 4(c) and show that the estimates presented in Table 8.20 are approximately:

$$\hat{\tau}_1 = \left(2500 - \hat{\lambda}_0\right)/\hat{\sigma} \qquad\qquad \hat{\tau}_1 = 0.329$$

$$\hat{\tau}_2 = \left(3000 - \hat{\lambda}_0\right)/\hat{\sigma} \qquad\qquad \hat{\tau}_2 = 1.592$$

$$\hat{\tau}_3 = \left(3500 - \hat{\lambda}_0\right)/\hat{\sigma} \qquad\qquad \hat{\tau}_3 = 2.854$$

$$\hat{\beta}_1 = \hat{\lambda}_1 \,/\, \hat{\sigma} \qquad\qquad\qquad \hat{\beta}_1 = 0.011$$

(e) By hand draw a facsimile of the density function shown in Figure 8.3 with the three vertical lines at the values 2500, 3000 and 3500. Using the results in equation (8.20), equation (8.21) and the estimates in Table 8.20 compute the value of the four areas under the hand drawn curve. Using these specific areas demonstrate that the relationship shown in equation (8.22) holds at each cutpoint.

There are four areas under the curve defined by the cutpoints, 2500, 3000 and 3500. These areas will be denoted a_1, a_2, a_3 and a_4 (from left to right).

Calculate a_1:

$$P\left[\varepsilon \le \tau_1 - 125 * \beta\right] = \frac{e^{\tau_1 - 125 * \beta}}{1 + e^{\tau_1 - 125 * \beta}} \qquad\qquad P\left[\varepsilon \le \tau_1 - 125 * \beta\right] = 0.260$$

Calculate a_2:

$$a_2 = \left(P\left[\varepsilon \le \tau_2 - 125 * \beta\right] - a_1\right) = 0.554 - 0.260 = 0.294 \qquad\qquad a_2 = 0.294$$

In a similar manner, a_3 and a_4 can be calculated with the following results:

$$a_3 = 0.260$$

$$a_4 = 0.186$$

Use these values to demonstrate the relation shown in equation (8.22):

$$\ln\left[\frac{P(BWT4N \le 0 \mid LWT = 125)}{P(BWT4N > 0 \mid LWT = 125)}\right] = \ln\left[\frac{0.260}{1 - 0.260}\right] = \tau_1 - 125 * \beta = -1.046$$

The same relation can be confirmed for each of the other cutpoints.

(f) Repeat problem 4(e) for LWT=135 and show by direct calculation using areas under the two curves that the relationship in equation (8.24) holds at each cutpoint.

Using LWT=135 and repeating the analysis above the following values for a_1, a_2, a_3 and a_4 can be obtained:

$$a_1 = 0.239$$

$$a_2 = 0.287$$

$$a_3 = 0.271$$

$$a_4 = 0.203$$

Equation (8.24) shows that in this example, comparing LWT=135 with LWT=125:

$$\ln\left[\frac{P(Y \le k \mid x_1)}{P(Y > k \mid x_1)}\right] - \ln\left[\frac{P(Y \le k \mid x_0)}{P(Y > k \mid x_0)}\right] = -\beta(x_1 - x_0) = (-0.011)(10) = -0.11$$

This same result can be obtained by direct calculation using the areas under the two curves at each cutpoint:

For cutpoint 1:

$$\ln\left[\frac{P(BWT4N \le 0 \mid LWT = 135)}{P(BWT4N > 0 \mid LWT = 135)}\right] - \ln\left[\frac{P(BWT4N \le 0 \mid LWT = 125)}{P(BWT4N > 0 \mid LWT = 125)}\right]$$
$$= \left(\frac{0.239}{1 - 0.239}\right) - \left(\frac{0.260}{1 - 0.260}\right) = -0.11$$

This can be shown for each of the other cutpoints using the values calculated for a_2, a_3 and a_4 at each value of LWT above.

5. *Using the data from the longitudinal low birth weight study and considering all the covariates (be sure to consider the possibility of interactions among the covariates):*

(a) Find the best cluster-specific and population average models.

The Longitudinal Low Birthweight data come from a hypothetical data set based on the low birth weight data described in Section 1.6.2. There are 188 women included in the data set. A hypothetical additional number of births was generated for each woman and varied between 1 and 3. As a result, the data set has information on 488 births.

The distribution of the outcome variable (LOW) is shown below:

```
. tab low

        low |      Freq.     Percent        Cum.
------------+-----------------------------------
          0 |        337       69.06       69.06
          1 |        151       30.94      100.00
------------+-----------------------------------
      Total |        488      100.00
```

Indicator variables for RACE are created (RACE_2 and RACE_3).

The remaining variables, AGE, LWT, and SMOKE were originally included in their original forms.

Since there are not very many covariates in this dataset, initial models were built including all four potential covariates.

Cluster-specific model

```
. xtlogit low age lwt smoke race_2 race_3, re i(id)

Fitting comparison model:

Iteration 0:   log likelihood = -301.89672
Iteration 1:   log likelihood = -282.78488
Iteration 2:   log likelihood = -282.50883
Iteration 3:   log likelihood = -282.50857

Fitting full model:

rho =  0.0     log likelihood = -282.50857
rho =  0.1     log likelihood = -277.80369
rho -  0.2     log likelihood = -272.95398
rho =  0.3     log likelihood = -267.96253
rho =  0.4     log likelihood = -262.83474
rho =  0.5     log likelihood =  -257.5847
rho =  0.6     log likelihood = -252.25383
rho =  0.7     log likelihood = -246.97109
rho =  0.8     log likelihood = -242.19389
Iteration 0:   log likelihood = -246.97109
Iteration 1:   log likelihood =  -238.4542
Iteration 2:   log likelihood = -235.87574
Iteration 3:   log likelihood = -233.91314
Iteration 4:   log likelihood = -231.15986
Iteration 5:   log likelihood = -230.86316
Iteration 6:   log likelihood = -230.85309
Iteration 7:   log likelihood = -230.85306
```

```
Random-effects logit                    Number of obs      =        488
Group variable (i) : id                 Number of groups   =        188

Random effects u_i ~ Gaussian           Obs per group: min =          2
                                                       avg =        2.6
                                                       max =          4

                                        Wald chi2(5)       =      12.17
Log likelihood  = -230.85306            Prob > chi2        =     0.0325

------------------------------------------------------------------------------
         low |      Coef.   Std. Err.      z    P>|z|     [95% Conf. Interval]
-------------+----------------------------------------------------------------
         age |   .1383036   .0525662     2.63   0.009     .0352756    .2413315
         lwt |  -.0252472   .0117643    -2.15   0.032    -.0483047   -.0021897
       smoke |   .8943653   .8259957     1.08   0.279    -.7245565    2.513287
      race_2 |   -1.75725   .9075702    -1.94   0.053    -3.536055    .0215552
      race_3 |   -1.99637    1.08761    -1.84   0.066    -4.128046    .1353065
       _cons |  -1.506111   1.673146    -0.90   0.368    -4.785416    1.773194
-------------+----------------------------------------------------------------
    /lnsig2u |   2.483271   .3025319                      1.890319    3.076223
-------------+----------------------------------------------------------------
     sigma_u |    3.46127   .5235724                      2.573224    4.655789
         rho |   .9229607   .0215113                      .8687919    .9559012
------------------------------------------------------------------------------
Likelihood ratio test of rho=0: chibar2(01) =    103.31 Prob >= chibar2 = 0.000
```

The Wald statistic was used to identify variables eligible for exclusion from the model. SMOKE was the first variable to be excluded from the model. Its impact on the coefficients for the other covariates in the model was assessed. Inclusion of this variable resulted in significant changes in the coefficients for RACE_3 (more than 30%). Since it appears that SMOKE is an important confounder, it will be included in the model. At this point, the remaining variables in the model: LWT, AGE, and RACE were found to be significantly associated with LOW. Therefore, all variables will be included in the model.

AGE and LWT are the only continuous variables in the model. The appropriateness of modeling these variables in their linear forms was assessed using the method of fractional polynomials (assuming that the observations are not correlated). The results (shown below) indicate that it is appropriate to use both covariates in their linear forms.

```
. fracpoly logit low age lwt smoke race_2 race_3 , compare
........
-> gen age_1 = x^3 if _sample
-> gen age_2 = x^3*ln(x) if _sample
where: x = age/10

Iteration 0:   Log Likelihood =-301.89672
Iteration 1:   Log Likelihood =-282.75983
Iteration 2:   Log Likelihood =-282.48372
Iteration 3:   Log Likelihood =-282.48347

Logit Estimates                             Number of obs =     488
                                            chi2(6)       =   38.83
                                            Prob > chi2   =  0.0000
Log Likelihood = -282.48347                 Pseudo R2     =  0.0643


---------------------------------------------------------------------
    low |     Coef.    Std. Err.      z     P>|z|    [95% Conf. Interval]
--------+------------------------------------------------------------
  age_1 |   .0565419    .0765831    0.738   0.460    -.0935582    .2066421
  age_2 |  -.0289432    .0527887   -0.548   0.583    -.1324072    .0745208
    lwt |  -.0093923    .0037513   -2.504   0.012    -.0167448   -.0020398
  smoke |    .541153    .2206954    2.452   0.014      .108598     .973708
 race_2 |  -.6938624    .3311255   -2.095   0.036    -1.342856   -.0448684
 race_3 |  -.8064714    .2523522   -3.196   0.001    -1.301073   -.3118701
  _cons |   .1010551    .6687954    0.151   0.880     -1.20976     1.41187
---------------------------------------------------------------------
Deviance:   564.967. Best powers of age among 44 models fit: 3 3.

Fractional Polynomial Model Comparisons:

------------------------------------------------------------------
age              df        Deviance      Gain    P(term) Powers
------------------------------------------------------------------
Not in model      0         568.976       --       --
Linear            1         565.017     0.000     0.047   1
m = 1             2         565.017     0.000     1.000   1
m = 2             4         564.967     0.050     0.975   3 3
------------------------------------------------------------------
```

```
. fracpoly logit low lwt age  smoke race_2 race_3 , compare
.......
-> gen lwt_1 = x if _sample
-> gen lwt_2 = x^3 if _sample
where: x = lwt/100

Iteration 0:  Log Likelihood =-301.89672
Iteration 1:  Log Likelihood =-282.12323
Iteration 2:  Log Likelihood =-281.46143
Iteration 3:  Log Likelihood =-281.44286
Iteration 4:  Log Likelihood =-281.44283

Logit Estimates                          Number of obs =     488
                                         chi2(6)       =   40.91
                                         Prob > chi2   =  0.0000
Log Likelihood = -281.44283              Pseudo R2     =  0.0678


------------------------------------------------------------------------------
    low |     Coef.   Std. Err.      z      P>|z|     [95% Conf. Interval]
--------+---------------------------------------------------------------------
  lwt_1 |  .9791524   1.424931     0.687   0.492     -1.81366    3.771965
  lwt_2 | -.2577502   .1894257    -1.361   0.174     -.6290179   .1135174
    age |  .0333357   .0191292     1.743   0.081     -.0041568   .0708282
  smoke |  .5813494   .2233819     2.602   0.009      .1435289   1.01917
 race_2 | -.656537    .3323909    -1.975   0.048     -1.308011  -.0050629
 race_3 | -.7796543   .2529301    -3.082   0.002     -1.275388  -.2839203
  _cons | -2.156066   1.445174    -1.492   0.136     -4.988555   .6764242
------------------------------------------------------------------------------
Deviance:  562.886. Best powers of lwt among 44 models fit: 1 3.

Fractional Polynomial Model Comparisons:
------------------------------------------------------------------
lwt            df    Deviance    Gain   P(term) Powers
------------------------------------------------------------------
Not in model    0    571.684     --      --
Linear          1    565.017    0.000   0.010   1
m = 1           2    563.376    1.641   0.200   3
m = 2           4    562.886    2.131   0.782   1 3
------------------------------------------------------------------
```

The final step in the model building process was an evaluation of interactions.
Interaction terms were created between each of the main effects in the model. There was
a significant interaction between AGE and SMOKE in the model. This interaction will be
included in the preliminary final cluster-specific model:

```
. xtlogit low age lwt race_2 race_3 smoke  smkage, re i(id)

Fitting comparison model:

Iteration 0:    log likelihood = -301.89672
Iteration 1:    log likelihood = -277.69028
Iteration 2:    log likelihood = -277.32848
Iteration 3:    log likelihood = -277.3281

Fitting full model:

rho =  0.0     log likelihood =  -277.3281
rho =  0.1     log likelihood =  -272.6071
rho =  0.2     log likelihood = -267.75376
rho =  0.3     log likelihood = -262.77413
rho =  0.4     log likelihood = -257.67566
rho =  0.5     log likelihood = -252.47405
rho =  0.6     log likelihood = -247.21218
rho =  0.7     log likelihood = -242.02086
rho =  0.8     log likelihood = -237.35301
Iteration 0:    log likelihood = -242.02086
Iteration 1:    log likelihood = -230.42025
Iteration 2:    log likelihood =  -227.2726
Iteration 3:    log likelihood =  -223.5933
Iteration 4:    log likelihood = -222.93512
Iteration 5:    log likelihood = -222.91785
Iteration 6:    log likelihood = -222.91758
Iteration 7:    log likelihood = -222.91758
```

| Random-effects logit | | | | Number of obs | = | 488 |
| Group variable (i) : id | | | | Number of groups | = | 188 |

Random effects u_i ~ Gaussian

				Obs per group: min =	2
				avg =	2.6
				max =	4

| | | | | Wald chi2(6) | = | 22.09 |
| Log likelihood = -222.91758 | | | | Prob > chi2 | = | 0.0012 |

low	Coef.	Std. Err.	z	P>|z|	[95% Conf. Interval]	
age	-.0282505	.0596404	-0.47	0.636	-.1451437	.0886426
lwt	-.0219773	.0112462	-1.95	0.051	-.0440194	.0000648
race_2	-2.604586	1.270108	-2.05	0.040	-5.093953	-.1152190
race_3	-2.170963	1.967463	-1.10	0.270	-6.02712	1.685195
smoke	-7.608599	2.657195	-2.86	0.004	-12.81661	-2.400592
smkage	.337908	.0943219	3.58	0.000	.1530404	.5227756
_cons	2.189793	2.416795	0.91	0.365	-2.547038	6.926624
/lnsig2u	2.66859	.3161403			2.048966	3.288214
sigma_u	3.797318	.6002426			2.785655	5.176384
rho	.9351476	.0191728			.8858431	.9640222

```
Likelihood ratio test of rho=0: chibar2(01) =   108.82 Prob >= chibar2 = 0.000
```

The model shown above indicates that when the SMKAGE interaction term is included in the cluster-specific model, there are numerical problems. Exploration reveals several zero cells. For this reason, the interaction term will not be included in the final model (shown below).

```
. xtlogit low age lwt smoke race_2 race_3, re i(id)

Fitting comparison model:

Iteration 0:    log likelihood = -301.89672
Iteration 1:    log likelihood = -282.78488
Iteration 2:    log likelihood = -282.50883
Iteration 3:    log likelihood = -282.50857

Fitting full model:

rho =  0.0      log likelihood = -282.50857
rho =  0.1      log likelihood = -277.80369
rho =  0.2      log likelihood = -272.95398
rho =  0.3      log likelihood = -267.96253
rho =  0.4      log likelihood = -262.83474
rho =  0.5      log likelihood =  -257.5847
rho =  0.6      log likelihood = -252.25383
rho =  0.7      log likelihood = -246.97109
rho =  0.8      log likelihood = -242.19389
Iteration 0:    log likelihood = -246.97109
Iteration 1:    log likelihood =  -238.4542
Iteration 2:    log likelihood = -235.87574
Iteration 3:    log likelihood = -233.91314
Iteration 4:    log likelihood = -231.15986
Iteration 5:    log likelihood = -230.86316
Iteration 6:    log likelihood = -230.85309
Iteration 7:    log likelihood = -230.85306
```

```
Random-effects logit                    Number of obs      =        488
Group variable (i) : id                 Number of groups   =        188

Random effects u_i ~ Gaussian           Obs per group: min =          2
                                                       avg =        2.6
                                                       max =          4

                                        Wald chi2(5)       =      12.17
Log likelihood  = -230.85306            Prob > chi2        =     0.0325
```

```
------------------------------------------------------------------------------
         low |     Coef.   Std. Err.      z    P>|z|    [95% Conf. Interval]
-------------+----------------------------------------------------------------
         age |   .1383036   .0525662     2.63   0.009     .0352756    .2413315
         lwt |  -.0252472   .0117643    -2.15   0.032    -.0483047   -.0021897
       smoke |   .8943653   .8259957     1.08   0.279    -.7245565    2.513287
      race_2 |   -1.75725   .9075702    -1.94   0.053    -3.536055    .0215552
      race_3 |   -1.99637    1.08761    -1.84   0.066    -4.128046    .1353065
       _cons |  -1.506111   1.673146    -0.90   0.368    -4.785416    1.773194
-------------+----------------------------------------------------------------
    /lnsig2u |   2.483271   .3025319                      1.890319    3.076223
-------------+----------------------------------------------------------------
     sigma_u |    3.46127   .5235724                      2.573224    4.655789
         rho |   .9229607   .0215113                      .8687919    .9559012
------------------------------------------------------------------------------
Likelihood ratio test of rho=0: chibar2(01) =    103.31 Prob >= chibar2 = 0.000
```

Population average model

```
.  xtlogit low age lwt smoke race_2 race_3, pa i(id) robust

Iteration 1: tolerance = .31034806
Iteration 2: tolerance = .00817963
Iteration 3: tolerance = .00024207
Iteration 4: tolerance = 5.612e-06
Iteration 5: tolerance = 1.241e-07

GEE population-averaged model              Number of obs      =        488
Group variable:                       id   Number of groups   =        188
Link:                              logit   Obs per group: min =          2
Family:                         binomial                  avg =        2.6
Correlation:                 exchangeable                 max =          4
                                           Wald chi2(5)       =      18.18
Scale parameter:                       1   Prob > chi2        =     0.0027

                              (standard errors adjusted for clustering on id)
-----------------------------------------------------------------------------
             |             Semi-robust
         low |      Coef.   Std. Err.      z    P>|z|     [95% Conf. Interval]
-------------+---------------------------------------------------------------
         age |   .0556856   .0202199     2.75   0.006     .0160554    .0953158
         lwt |  -.0096687   .0043689    -2.21   0.027    -.0182316   -.0011058
       smoke |   .4731122   .3152191     1.50   0.133    -.1447058     1.09093
      race_2 |  -.6417466   .4423741    -1.45   0.147    -1.508784    .2252906
      race_3 |  -.7241344   .3615349    -2.00   0.045     -1.43273   -.0155391
       _cons |  -.7790307   .6893797    -1.13   0.258     -2.13019    .5721288
-----------------------------------------------------------------------------
```

The Wald statistic was used to identify variables eligible for exclusion from the model. SMOKE was the first variable to be excluded from the model. Its impact on the coefficients for the other covariates in the model was assessed. Inclusion of this variable resulted in significant changes in the coefficients for RACE_3 (more than 25%). Since it appears that SMOKE is an important confounder, it will be included in the model. At this point, the remaining variables in the model: LWT, AGE, and RACE were found to be significantly associated with LOW. Therefore, all variables will be included in the model.

AGE and LWT are the only continuous variables in the model. The appropriateness of modeling these variables in their linear forms was assessed using the method of fractional polynomials (assuming that the observations are not correlated). Since the fractional polynomial analysis is identical to that done for the cluster specific model we do not repeat the output. The results showed that the model is not significantly non-linear in age or lwt.

The final step in the model building process was an evaluation of interactions. Interaction terms were created between each of the main effects in the model. There was a significant interaction between AGE and SMOKE in the model. After fitting the model the coefficient for SMOKE was quite large and gave an indication of some numerical instability. Thus we decided to leave it out. The resulting model is also the same as the cluster-specific model, making subsequent cross-model comparisons a lot easier

(b) Evaluate the fit of the two models obtained in problem 5(a).

The fit of the preliminary final models from 5(a) can be evaluated using the Hosmer-Lemeshow test with the beta option in STATA. Following the fit of the preliminary final models, a matrix of coefficients is created and used for the goodness-of-fit test.

Cluster-Specific

In order for the Hosmer-Lemeshow test to be used with the cluster-specific model one must use fitted values that include an estimate of the random effect term, α_i, as well as the regression coefficients, as shown on page 328 of the text.

The difficulty is that STATA does not provide estimates of the random effect terms; but SAS does. However the estimates of the coefficients in STATA and SAS differ slightly. Following the procedure in the text we evaluate the fit of the model obtained by SAS, shown in the following output.

```
The NLMIXED Procedure

                        Specifications

Data Set                             A.CLSLOWBWT
Dependent Variable                   low
Distribution for Dependent Variable  Binomial
Random Effects                       u
Distribution for Random Effects      Normal
Subject Variable                     id
Optimization Technique               Dual Quasi-Newton
Integration Method                   Adaptive Gaussian
                                     Quadrature


                Dimensions

Observations Used             488
Observations Not Used           0
Total Observations            488
Subjects                      188
Max Obs Per Subject             4
Parameters                      7
Quadrature Points              21


                           Parameters

    b0       bage      blwt     bsmoke       br2       br3      s2u   NegLogLike

 -1.5061    0.1383   -0.0252    0.8943    -0.1757    -1.996    15.5   232.062778
```

```
                           Iteration History

 Iter      Calls    NegLogLike        Diff     MaxGrad        Slope

    1         5    231.546512     0.516266    6.018698     -2107.2
    2         7    231.50353      0.042982    1.282362     -0.43926
    3        10    230.467118     1.036412   40.38534      -0.02483
    4        11    230.428123     0.038995    2.668757     -0.07135
    5        13    230.41577      0.012353   22.89535      -0.00448
    6        15    230.312627     0.103143    2.629536     -0.01819
    7        17    230.312163     0.000464    2.888576     -0.00044
    8        19    230.304732     0.007431    0.223911     -0.00045
    9        21    230.303874     0.000859    0.10369      -0.00152
   10        23    230.290442     0.013431    0.798599      -0.0002
   11        25    230.287644     0.002798    0.066346     -0.00548
   12        27    230.287491     0.000152    0.276511     -0.00006
   13        31    230.257927     0.029564    2.916898     -0.00024
   14        33    230.253869     0.004058    0.061739     -0.00849
   15        35    230.25384      0.000029    0.02736      -0.00006
 The SAS System                                                          4

 The NLMIXED Procedure

                           Iteration History

 Iter      Calls    NegLogLike        Diff     MaxGrad        Slope

   16        37    230.25384     1.341E-7     0.005593     -2.83E-7


        NOTE: GCONV convergence criterion satisfied.


           Fit Statistics

-2 Log Likelihood                  460.5
AIC (smaller is better)            474.5
AICC (smaller is better)           474.7
BIC (smaller is better)            497.2
```

```
                             Parameter Estimates

                  Standard
Parameter  Estimate    Error    DF  t Value   Pr > |t|  Alpha    Lower      Upper    Gradient

b0         -2.1032    1.6605    187   -1.27    0.2069    0.05   -5.3790     1.1725   0.000047
bage        0.1371    0.05296   187    2.59    0.0104    0.05    0.03268    0.2416   0.000909
blwt       -0.02258   0.01157   187   -1.95    0.0524    0.05   -0.04540   0.000232  0.005593
bsmoke      1.1892    0.7675    187    1.55    0.1230    0.05   -0.3249     2.7034   9.758E-6
br2        -1.7353    1.1432    187   -1.52    0.1307    0.05   -3.9904     0.5199   4.774E-6
br3        -1.8709    0.8677    187   -2.16    0.0324    0.05   -3.5828    -0.1591  -1.07E-6
s2u        13.7943    4.7339    187    2.91    0.0040    0.05    4.4556    23.1329   5.18E-6
```

We evaluated the fit of this model using the same procedure described in the text. The key steps are creating an output file in SAS containing the estimates of the random effects and the model dependent variable, LOW, and the model covariates. These data were used to create a STATA file. Fit was evaluated as follows:

```
. mat input  b=(0.1371 -0.02258 1.1992  1.7353 -1.8709 1 -2.1032)

. matrix colnames b=age lwt smoke   race_2 race_3 u _cons

. quietly logit low  age lwt smoke race_2 race_3,offset(u)
```

Fit using the coefficients from the SAS model

```
. lfit low, beta(b) group(10) table

Logistic model for low, goodness-of-fit test
(Table collapsed on quantiles of estimated probabilities)
```

_Group	_Prob	_Obs_1	_Exp_1	_Obs_0	_Exp_0	_Total
1	0.0114	0	0.4	49	48.6	49
2	0.0167	0	0.7	49	48.3	49
3	0.0249	0	1.0	49	48.0	49
4	0.0381	0	1.5	49	47.5	49
5	0.0553	0	2.3	48	45.7	48
6	0.2682	1	5.8	48	43.2	49
7	0.4497	20	18.1	29	30.9	49
8	0.7809	33	28.7	16	20.3	49
9	0.8886	49	41.9	0	7.1	49
10	0.9793	48	44.6	0	3.4	48

```
        number of observations =       488
               number of groups =        10
    Hosmer-Lemeshow chi2(8)    =     24.42
            Prob > chi2 =          0.0019
```

Since the p-value for the goodness-of-fit test is quite small, we have evidence of lack of model fit. However, 6 of the 20 cells have expected values less than 5 and 3 cells have expected values of 1.0 or smaller. Thus the p-value may not be accurate. Visual inspection of the observed and expected counts in the 2 by 10 table above show reasonable agreement indicating the fit may be better than indicated by the p-value for the test.

Population Average

```
. matrix input b=get(_b)
```

```
. lfit low, group (10) table beta(b)

Logistic model for low, goodness-of-fit test
(Table collapsed on quantiles of estimated probabilities)
```

_Group	_Prob	_Obs_1	_Exp_1	_Obs_0	_Exp_0	_Total
1	0.1662	6	6.6	43	42.4	49
2	0.1960	11	8.9	38	40.1	49
3	0.2243	4	10.3	45	38.7	49
4	0.2608	12	11.7	37	37.3	49
5	0.2965	13	13.3	35	34.7	48
6	0.3335	17	15.5	32	33.5	49
7	0.3781	18	17.5	31	31.5	49
8	0.4293	27	19.8	22	29.2	49
9	0.5070	20	23.1	29	25.9	49
10	0.6624	23	26.8	25	21.2	48

```
        number of observations =       488
               number of groups =        10
    Hosmer-Lemeshow chi2(8)    =     12.27
            Prob > chi2 =          0.1395
```

Since the p-value is 0.14 we have evidence that the population average model fits.

(c) Prepare a separate table for each model obtained in 5(a) containing estimates of the odds ratios with 95 percent confidence intervals.

Cluster-Specific

Variable	Coefficient	Standard Error	Odds Ratio, 95% Confidence Interval
Smoking During Pregnancy (vs. No smoking)	0.894	0.8260	2.45 (0.48, 12.35)
Black Race (vs. White)	-1.757	0.9076	0.17 (0.03, 1.02)
Other Race (vs. White)	-1.996	1.0876	0.14 (0.02, 1.14)
Weight of Mother at Last Menstrual Period (pounds)	-0.025	0.0118	0.78* (0.62, 0.98)
Age (years)	0.138	0.0526	2.00** (1.19, 3.34)

*Odds ratio for a 10-pound increase in LWT
**Odds ratio for a 5-year increase in AGE.

Population Average

Variable	Coefficient	Standard Error	Odds Ratio, 95% Confidence Interval
Smoking During Pregnancy (vs. No smoking)	0.473	0.3152	1.60 (0.87, 2.98)
Black Race (vs. White)	-0.642	0.4424	0.53 (0.22, 1.25)
Other Race (vs. White)	-0.724	0.3615	0.48 (0.24, 0.98)
Weight of Mother at Last Menstrual Period (pounds)	-0.010	0.0044	0.91* (0.83, 0.99)
Age (years)	0.056	0.0202	1.32** (1.08, 1.61)

*Odds ratio for a 10-pound increase in LWT
**Odds ratio for a 5-year increase in AGE.

(d) Compare the interpretation of the point estimates of the odds ratios from the cluster-specific model and population average model.

The interpretation of the odds ratios from a fitted cluster-specific model applies to subjects with a common, but unobserved value of the random effect α_{ii}. In the model shown above, this means that by smoking during pregnancy a woman has increased her odds of a low birthweight baby by 2.45 over the odds if she did not smoke holding her age, weight and race constant. Also, the cluster-specfic odds ratio for AGE indicates that the odds of a woman having a low birthweight baby in five years is 2.00 times the odds at her current age holding weight, smoking status and race constant. The cluster-specific odds ratio for LWT indicates that the odds of a woman having a low birthweight baby if she gained 10 pounds is 22% less than the odds at her current weight, holding age at pregnancy and smoking status constant. It is more difficult to interpret the odds ratios for non-white race as race is not a modifiable factor. The odds ratios suggest that the odds of hypothetical groups of women with the same value of the random effect α_i, having a low birthweight baby who are non-white are more than 80% less than the odds of white women with this same random effect, holding age at pregnancy, weight and smoking status constant.

The interpretation of the odds ratios for the fitted population average models are easier to interpret. The odds of a low birthweight baby computed from the proportion of women who smoke is 1.60 times as high as that based on the proportion of women who do not smoke, holding age, weight and race constant. The population average odds ratio for a five-year increase in age at delivery is 1.32, meaning that the odds of low birth weight computed from the proportion of women who are five years older than some reference level for age is 1.32 times higher than that based on the proportion of women at the reference age, holding smoking status, race and weight constant. The odds ratios for race and weight can be interpreted in a similar fashion. This interpretation does not use the information that is available in the repeated measurements of the covariates on study subjects.

6. *Using the cluster-specific and population average models obtained in Exercise 5(a) explore alternative ways of including the weight of the mother at the last menstrual period. For example, one alternative is to use the weight at the first birth as a cluster level covariate. Other representations are possible. For each alternative fit the cluster-specific and population average model, estimate an odds ratio for weight and compare their interpretation.*

Cluster-specific model

```
. xtlogit low  smoke race_2 race_3 age firstlwt, re i(id)

Fitting comparison model:

Iteration 0:    log likelihood = -301.89672
Iteration 1:    log likelihood = -282.38013
Iteration 2:    log likelihood = -282.08673
Iteration 3:    log likelihood = -282.08641
Iteration 4:    log likelihood = -282.08641

Fitting full model:

rho = 0.0       log likelihood = -282.08641
rho = 0.1       log likelihood = -277.38951
rho = 0.2       log likelihood =  -272.5425
rho = 0.3       log likelihood = -267.54933
rho = 0.4       log likelihood = -262.41623
rho = 0.5       log likelihood = -257.15817
rho = 0.6       log likelihood = -251.81751
rho = 0.7       log likelihood = -246.52467
rho = 0.8       log likelihood = -241.74117
Iteration 0:    log likelihood = -246.52467
Iteration 1:    log likelihood = -238.26235
Iteration 2:    log likelihood = -234.05613
Iteration 3:    log likelihood = -232.29012
Iteration 4:    log likelihood = -231.53338
Iteration 5:    log likelihood = -230.65682
Iteration 6:    log likelihood = -230.62042
Iteration 7:    log likelihood = -230.62024
Iteration 8:    log likelihood = -230.62024


Random-effects logit                     Number of obs      =        488
Group variable (i) : id                  Number of groups   =        188

Random effects u_i ~ Gaussian            Obs per group: min =          2
                                                        avg =        2.6
                                                        max =          4

                                         Wald chi2(5)       =      10.98
Log likelihood  = -230.62024             Prob > chi2        =     0.0518

------------------------------------------------------------------------------
         low |      Coef.   Std. Err.      z    P>|z|     [95% Conf. Interval]
-------------+----------------------------------------------------------------
       smoke |   .9059917   .7904939     1.15   0.252    -.6433478    2.455331
      race_2 |  -1.699181   .8686473    -1.96   0.050    -3.401699     .003336
      race_3 |  -2.097396   1.083934    -1.93   0.053    -4.221867    .0270748
         age |   .0888412    .042723     2.08   0.038     .0051056    .1725768
    firstlwt |  -.0294358   .0138145    -2.13   0.033    -.0565118   -.0023598
       _cons |   .0269182   2.094778     0.01   0.990    -4.078772    4.132608
-------------+----------------------------------------------------------------
    /lnsig2u |   2.452329   .2920272                      1.879967    3.024692
-------------+----------------------------------------------------------------
     sigma_u |   3.408133   .4976338                      2.559939    4.537363
         rho |   .9207316   .0213136                      .8676073    .9536773
------------------------------------------------------------------------------
Likelihood ratio test of rho=0: chibar2(01) =    102.93 Prob >= chibar2 = 0.000
```

The interpretation of the odds ratio for FIRSTLWT (or weight of the mother at the last menstrual period before her first pregnancy) from the fitted cluster-specific model applies to subjects with a common, but unobserved value of the random effect α_i. In the model shown above, this means the odds of a woman having a low birthweight baby if she had originally been 10 pounds heavier prior to her first pregnancy would have been 25% less than the odds at her actual weight, holding age at pregnancy, smoking status and race constant.

Population average model

```
. xtlogit low  smoke race_2 race_3 age firstlwt, pa i(id) robust

Iteration 1: tolerance = .05507063
Iteration 2: tolerance = .00405163
Iteration 3: tolerance = .00005178
Iteration 4: tolerance = 2.598e-06
Iteration 5: tolerance = 5.381e-08

GEE population-averaged model              Number of obs      =        488
Group variable:                        id  Number of groups   =        188
Link:                               logit  Obs per group: min =          2
Family:                          binomial                 avg =        2.6
Correlation:                  exchangeable                 max =          4
                                           Wald chi2(5)       =      18.13
Scale parameter:                        1  Prob > chi2        =     0.0028

                             (standard errors adjusted for clustering on id)
------------------------------------------------------------------------------
             |             Semi-robust
         low |      Coef.   Std. Err.      z    P>|z|     [95% Conf. Interval]
-------------+----------------------------------------------------------------
       smoke |   .4390475   .3196164     1.37   0.170    -.1873892    1.065484
      race_2 |  -.6782924   .4458803    -1.52   0.128    -1.552202     .195617
      race_3 |   -.785503   .3727602    -2.11   0.035      -1.5161   -.0549064
         age |   .0380456   .0173016     2.20   0.028     .0041351    .0719561
     firstlwt |  -.0119074   .0053511    -2.23   0.026    -.0223955   -.0014194
       _cons |  -.1175686   .8993043    -0.13   0.896    -1.880173    1.645035
------------------------------------------------------------------------------
```

The odds of a low birthweight baby computed from the proportion of women who were 10 pounds heavier than some reference level for weight at the last menstrual period prior to the first pregnancy is 0.89 times as high as that based on the proportion of women at that reference weight, holding age, smoking status and race constant.

7. *Using the covariates in the population average model obtained in Exercise 5(a) explore alternative ways of including history of a low birth weight baby. For each model compute and interpret the estimate of the odds ratio and a 95% confidence interval for the previous low birth weight covariate. Recall that we fit these models using the usual logistic regression model.*

(a) *Fit the model that adds the outcome of the previous birth to the model. In this problem explore the use of two versions of this covariate: one that assigns a missing value for the first birth and one that assigns a value of zero to the first birth.*

Missing value for First Birth

```
. logit low age lwt race_2 race_3 smoke prevlow

Iteration 0:    log likelihood = -189.53785
Iteration 1:    log likelihood = -118.17303
Iteration 2:    log likelihood = -113.59412
Iteration 3:    log likelihood = -113.31624
Iteration 4:    log likelihood = -113.31447
Iteration 5:    log likelihood = -113.31447

Logit estimates                           Number of obs   =        300
                                          LR chi2(6)      =     152.45
                                          Prob > chi2     =     0.0000
Log likelihood = -113.31447               Pseudo R2       =     0.4022


------------------------------------------------------------------------------
     low |      Coef.   Std. Err.      z    P>|z|     [95% Conf. Interval]
---------+--------------------------------------------------------------------
     age |   .0729793   .0343062     2.13   0.033     .0057403    .1402102
     lwt |  -.0172453   .0068763    -2.51   0.012    -.0307226   -.0037679
  race_2 |  -.3186634   .5543462    -0.57   0.565    -1.405162    .7678352
  race_3 |  -.4682124   .4243445    -1.10   0.270    -1.299912    .3634875
   smoke |   1.520545   .3910659     3.89   0.000     .7540697     2.28702
 prevlow |     3.3471   .3896719     8.59   0.000     2.583357    4.110843
   _cons |  -1.917103   1.358064    -1.41   0.158     -4.57886    .7446534
------------------------------------------------------------------------------
```

The odds ratio for PREVLOW is 28.27 (95% CI: 13.04, 61.27). The interpretation is that the odds of a low weight birth among women whose previous pregnancy resulted in a low weight birth is more than 28 times the odds for women whose previous pregnancy did not result in a low weight baby, controlling for age, weight, smoking status and race. The confidence interval suggests that the increased odds may be as low as 13 times or as high as 61 times with 95% confidence.

Zero value for First Birth

```
. logit low age lwt race_2 race_3 smoke prevlow

Iteration 0:    log likelihood = -301.89672
Iteration 1:    log likelihood = -236.38046
Iteration 2:    log likelihood = -234.83835
Iteration 3:    log likelihood = -234.82378
Iteration 4:    log likelihood = -234.82377

Logit estimates                           Number of obs   =        488
                                          LR chi2(6)      =     134.15
                                          Prob > chi2     =     0.0000
Log likelihood = -234.82377               Pseudo R2       =     0.2222

------------------------------------------------------------------------
      low |     Coef.    Std. Err.      z     P>|z|    [95% Conf. Interval]
----------+-------------------------------------------------------------
      age |   .0058856   .0217579    0.27    0.787   -.0367592    .0485304
      lwt |  -.0132877   .0044049   -3.02    0.003   -.0219213   -.0046542
   race_2 |  -.5826902   .3696292   -1.58    0.115    -1.30715    .1417697
   race_3 |  -.7804735    .284212   -2.75    0.006   -1.337519   -.2234282
    smoke |   .4836392   .2496511    1.94    0.053    -.005668    .9729464
  prevlow |   2.733602   .3154146    8.67    0.000    2.115401    3.351804
    _cons |   .4818496   .7402844    0.65    0.515   -.9690812     1.93278
------------------------------------------------------------------------
```

When the first birth is included (with a value of zero), the odds ratio for PREVLOW is 15.39 (95% CI: 8.29, 28.55). The interpretation is that the odds of a low weight birth among women whose previous pregnancy resulted in a low weight birth is more than 15 times the odds for women whose previous pregnancy did not result in a low weight baby, controlling for age, weight, smoking status and race. The confidence interval suggests that the increased odds may be as low as 8 times or as high as 29 times with 95% confidence.

(b) Fit the model that includes a dichotomous covariate indicating if any previous birth was of low weight. In this problem explore the use of two versions of this covariate: one that assigns a missing value for the first birth and one that assigns a value of zero to the first birth.

Missing value for First Birth

```
. logit low age lwt race_2 race_3 smoke everlow

Iteration 0:   log likelihood = -189.53785
Iteration 1:   log likelihood = -102.60387
Iteration 2:   log likelihood = -95.631717
Iteration 3:   log likelihood = -94.887364
Iteration 4:   log likelihood = -94.872044
Iteration 5:   log likelihood = -94.872035

Logit estimates                            Number of obs   =        300
                                           LR chi2(6)      =     189.33
                                           Prob > chi2     =     0.0000
Log likelihood = -94.872035                Pseudo R2       =     0.4995

------------------------------------------------------------------------------
     low |     Coef.   Std. Err.      z    P>|z|    [95% Conf. Interval]
---------+--------------------------------------------------------------------
     age |   .1009083   .0392876     2.57   0.010     .023906    .1779105
     lwt |  -.0206104   .0080671    -2.55   0.011    -.0364215   -.0047992
  race_2 |  -.2780634   .6474759    -0.43   0.668    -1.547093    .9909661
  race_3 |  -.3223043   .4690428    -0.69   0.492    -1.241611    .5970027
   smoke |   1.442997    .430752     3.35   0.001     .5987387   2.287255
 everlow |   3.976095   .4248813     9.36   0.000     3.143343   4.808847
   _cons |  -2.768711   1.532423    -1.81   0.071    -5.772205    .2347836
------------------------------------------------------------------------------
```

The odds ratio for EVERLOW is 53.31 (95% CI: 23.18, 122.59). The interpretation is that the odds of a low weight birth among women who have ever had a pregnancy resulting in a low weight birth is more than 53 times the odds for women with no previous pregnancies resulting a low weight baby, controlling for age, weight, smoking status and race. The confidence interval suggests that the increased odds may be as low as 23 times or as high as 123 times with 95% confidence.

Zero value for First Birth

```
. logit low age lwt race_2 race_3 smoke everlow

Iteration 0:    log likelihood = -301.89672
Iteration 1:    log likelihood = -222.73678
Iteration 2:    log likelihood = -220.47314
Iteration 3:    log likelihood = -220.43646
Iteration 4:    log likelihood = -220.43645

Logit estimates                          Number of obs   =        488
                                         LR chi2(6)      =     162.92
                                         Prob > chi2     =     0.0000
Log likelihood = -220.43645              Pseudo R2       =     0.2698

------------------------------------------------------------------------------
     low |      Coef.   Std. Err.       z    P>|z|     [95% Conf. Interval]
---------+--------------------------------------------------------------------
     age | -.0012162   .0228666    -0.05   0.958    -.046034    .0436015
     lwt | -.0151396    .004695    -3.22   0.001   -.0243417   -.0059376
  race_2 | -.5913913   .3847732    -1.54   0.124   -1.345533    .1627503
  race_3 | -.7928699   .2950917    -2.69   0.007   -1.371239   -.2145008
   smoke |  .3516766   .2603213     1.35   0.177   -.1585439    .861897
 everlow |  3.038634   .3155111     9.63   0.000    2.420244    3.657025
   _cons |   .826485    .774385     1.07   0.286   -.6912817    2.344252
------------------------------------------------------------------------------
```

When the first birth is included (with a value of zero), the odds ratio for EVERLOW is 20.88 (95% CI: 11.25, 38.75). The interpretation is that the odds of a low weight birth women who have ever had a pregnancy resulting in a low weight birth is more than 20 times the odds for women with no previous pregnancies resulting a low weight baby, controlling for age, weight, smoking status and race. The confidence interval suggests that the increased odds may be as low as 11 times or as high as 38 times with 95% confidence.

8. *Using the data from the ICU Study described in Chapter 1, Section 1.6.1, attempt to fit the usual logistic regression model containing type of admission (TYP) using subjects 25 years of age or younger. Why does the usual MLE have problems in this example? Fit the exact logistic regression model. Compute the point and 95% confidence interval estimates of the odds ratio.*

There are 27 subjects in the ICU data set who are 25 years of age or younger. When the usual logistic regression of STA on TYP is run on this subgroup of patients, the following results are obtained:

```
. logit sta typ

Note: typ~=1 predicts failure perfectly
      typ dropped and 3 obs not used

Iteration 0:  Log Likelihood =-6.8840636

Logit Estimates                                 Number of obs =      24
                                                chi2(0)       =    0.00
                                                Prob > chi2   =       .
Log Likelihood = -6.8840636                     Pseudo R2     = 0.0000


------------------------------------------------------------------------
    sta |     Coef.   Std. Err.       z     P>|z|    [95% Conf. Interval]
--------+---------------------------------------------------------------
  _cons | -2.397895   .7385489    -3.247    0.001    -3.845425   -.9503659
------------------------------------------------------------------------
```

The variable, TYP, our covariate of interest, is dropped from the model, because it perfectly predicts the outcome. This is due to a zero cell as shown below.

```
. tab sta typ

        |     typ
    sta |      0          1 |     Total
--------+----------------------+----------
      0 |      3         22 |        25
      1 |      0          2 |         2
--------+----------------------+----------
  Total |      3         24 |        27
```

The exact logistic regression model, obtained from LogXact is shown below.

```
=================================================================================
Binary Logistic Regression                           LogXact 4.1 for Windows
=================================================================================
Basic Information
Data file name          C:\logexact\icu.cyl
Model                   sta=%CONST+typ
Weight variable         Not specified
Stratum variable        <Unstratified>
Analysis type           Estimate : Exact
=================================================================================
Number of terms         2
Total observations      27
Observations rejected   173
Number of groups        2
=================================================================================
Summary statistics
=================================================================================
                        Statistic                    Value    DF     P-value
                        Deviance                     NA       NA     NA
                        Likelihood Ratio             NA       NA     NA
=================================================================================
Parameter Estimates
=================================================================================
              Point Estimate            Confidence interval and P-value for Beta
       Type   Beta     SE(Beta)    Type       95%      C.I.           Pvalue
                                               Lower    Upper          2*1-sided
=================================================================================
typ    MLE    ?        ?           Asymptotic ?        ?              ?
       MUE    -1.2126  NA          Exact      -3.8937  +INF           1.0000
%CONST MLE    ?        ?           Asymptotic ?        ?              ?
=================================================================================
Analysis time :00:00:01
Maximum Memory Used (KB) = 104
Optimum size of memory to solve the problem (KB) = 156
=================================================================================
```

The odds ratio can be obtained by exponentiating the exact estimate of beta, shown in bold above. The confidence interval can be obtained by exponentiating the endpoints of the confidence interval for beta above.

$$OR = \exp(-1.2126) = 0.30$$

95% Confidence Interval: $(0.02, \infty)$

9. *Repeat Exercise 8, fitting models containing systolic blood pressure (SYS).*

```
. logit sta sys

Iteration 0:  Log Likelihood =-7.1294054
Iteration 1:  Log Likelihood =-6.2880242
Iteration 2:  Log Likelihood =-6.1187967
Iteration 3:  Log Likelihood =-6.1128206
Iteration 4:  Log Likelihood =-6.1127998

Logit Estimates                          Number of obs =       27
                                         chi2(1)       =     2.03
                                         Prob > chi2   =   0.1539
Log Likelihood = -6.1127998              Pseudo R2     =   0.1426

------------------------------------------------------------------------------
     sta |     Coef.   Std. Err.       z     P>|z|     [95% Conf. Interval]
---------+--------------------------------------------------------------------
     sys |   .067887   .0541617     1.253    0.210    -.0382679     .174042
   _cons | -11.73796   7.745576    -1.515    0.130    -26.91901    3.443089
------------------------------------------------------------------------------
```

```
. tab sys sta

           | sta
       sys |         0          1 |     Total
-----------+----------------------+----------
       100 |         2          0 |         2
       104 |         3          0 |         3
       110 |         1          0 |         1
       112 |         2          0 |         2
       120 |         4          0 |         4
       130 |         3          0 |         3
       131 |         1          0 |         1
       136 |         1          0 |         1
       140 |         2          1 |         3
       142 |         2          0 |         2
       144 |         1          0 |         1
       146 |         1          0 |         1
       148 |         0          1 |         1
       156 |         1          0 |         1
       164 |         1          0 |         1
-----------+----------------------+----------
     Total |        25          2 |        27
```

```
========================================================================
Binary Logistic Regression                      LogXact 4.1 for Windows
========================================================================
Basic Information
Data file name          C:\hozstat\logexact\icu.cyl
Model                   sta=%CONST+sys
Weight variable         Not specified
Stratum variable        <Unstratified>
Analysis type           Estimate : Exact
========================================================================
Number of terms         2
Total observations      27
Observations rejected   173
Number of groups        15
========================================================================
Summary statistics
========================================================================
                  Statistic              Value     DF      P-value
                  Deviance               8.4065    13      0.8161
                  Likelihood Ratio       25.2043   2       0.0000
========================================================================
Parameter Estimates
========================================================================
              Point Estimate      Confidence interval and P-value for Beta
        Type  Beta    SE(Beta)  Type       95%     C.I.        Pvalue
                                          Lower     Upper      2*1-sided
========================================================================
sys     MLE   0.0679  0.0542   Asymptotic -0.0383  0.1740      0.2101
        CMLE  0.0639  0.0513   Exact      -0.0267  0.1905      0.1880
%CONST  MLE  -11.7380 7.7457   Asymptotic -26.9192 3.4433      0.1297
========================================================================
Analysis time :00:00:01
Maximum Memory Used (KB) = 108
Optimum size of memory to solve the problem (KB) = 162
========================================================================
```

OR_{exact}(10-point increase in SYS) = $\exp(0.639) = 1.89$

Exact 95% CI: (0.77, 6.72)

10. Evaluate the fit of the usual and exact logistic regression models shown in Table 8.36.

Usual Logistic Regression Model:

```
. logit low lwt smoke ptld

Iteration 0:   Log Likelihood =-40.607858
Iteration 1:   Log Likelihood =-35.561288
Iteration 2:   Log Likelihood =-35.279999
Iteration 3:   Log Likelihood =-35.276277
Iteration 4:   Log Likelihood =-35.276275

Logit Estimates                              Number of obs  =       69
                                             chi2(3)        =    10.66
                                             Prob > chi2    =   0.0137
Log Likelihood = -35.276275                  Pseudo R2      =   0.1313

------------------------------------------------------------------------
    low  |    Coef.    Std. Err.      z     P>|z|   [95% Conf. Interval]
---------+--------------------------------------------------------------
    lwt  | -.0193155   .011657    -1.657   0.098   -.0421628     .0035317
  smoke  |  .2489932   .6086555    0.409   0.682   -.9439497    1.441936
   ptld  |  1.392705   .6687295    2.083   0.037    .0820109    2.703391
  _cons  |  1.096696   1.559898    0.703   0.482   -1.960647     4.15404
------------------------------------------------------------------------
```

```
. lfit, group(10) table

Logistic model for low, goodness-of-fit test
(Table collapsed on quantiles of estimated probabilities)

_Group    _Prob     _Obs_1     _Exp_1    _Obs_0     _Exp_0     _Total
    1     0.0925        1        0.4         6         6.6        7
    2     0.1304        1        0.8         6         6.2        7
    3     0.1748        1        1.1         6         5.9        7
    4     0.1956        2        1.5         6         6.5        8
    5     0.2277        0        1.8         8         6.2        8
    6     0.2593        1        1.2         4         3.8        5
    7     0.3113        0        2.0         7         5.0        7
    8     0.3810        5        2.5         2         4.5        7
    9     0.6041        3        3.8         4         3.2        7
   10     0.7117        5        3.9         1         2.1        6

           number of observations =        69
              number of groups =           10
      Hosmer-Lemeshow chi2(8) =         11.35
                    Prob > chi2 =       0.1827
```

Exact Logistic Regression Model:

Capability not currently available with existing LogXact program.

However we can use STATA by inputting the coefficients.

```
. mat input  b=(-0.018 0.256 1.310 0.331)

. matrix colnames b= lwt smoke  ptd  _cons

. mat list b

b[1,4]
        lwt   smoke    ptd   _cons
r1   -.018    .256    1.31   .331

. lfit low, beta(b) group(10) table

Logistic model for low, goodness-of-fit test
(Table collapsed on quantiles of estimated probabilities)

_Group      _Prob      _Obs_1     _Exp_1    _Obs_0     _Exp_0      _Total
    1      0.0563         1         0.3        6         6.7          7
    2      0.0801         1         0.5        6         6.5          7
    3      0.1057         1         0.6        6         6.4          7
    4      0.1183         2         0.9        6         7.1          8
    5      0.1384         0         1.1        8         6.9          8
    6      0.1612         1         0.8        4         4.2          5
    7      0.1948         0         1.2        7         5.8          7
    8      0.2455         5         1.6        2         5.4          7
    9      0.4346         3         2.6        4         4.4          7
   10      0.5466         5         2.9        1         3.1          6

            number of observations =          69
                  number of  groups =          10
          Hosmer-Lemeshow chi2(8)  =       19.76
                   Prob > chi2  =        0.0113
```

11. *Consider the low birth weight study. What sample size would be needed in a new study to be able to detect that the odds of a low birthweight baby among women who smoke during pregnancy is 2.5 times that of women who do not smoke, using an $\alpha = 0.05$ type I error probability and $1 - \theta = 0.80$ power?*

Use Equation 8.44 to calculate the necessary sample size.

$$n = \left(1 + 2P_0\right) * \frac{\left(z_{1-\alpha}\sqrt{\dfrac{1}{1-\pi} + \dfrac{1}{\pi}} + z_{1-\theta}\sqrt{\dfrac{1}{1-\pi} + \dfrac{1}{\pi e^{\beta_1^*}}}\right)^2}{P_0\beta_1^{*2}}$$

where, P_0 is the response probability for non-smokers. This can be obtained using the cross-classification of LOW by SMOKE shown below.

```
.  tab low smoke, row col chi2

            |      smoke
       low  |        0           1  |     Total
    --------+----------------------+----------
         0  |       86          44  |       130
            |    66.15       33.85  |    100.00
            |    74.78       59.46  |     68.78
    --------+----------------------+----------
         1  |       29          30  |        59
            |    49.15       50.85  |    100.00
            |    25.22       40.54  |     31.22
    --------+----------------------+----------
     Total  |      115          74  |       189
            |    60.85       39.15  |    100.00
            |   100.00      100.00  |    100.00

          Pearson chi2(1)  =     4.9237    Pr = 0.026
```

P_0 can also be obtained using Equation 8.47 and the results from the univariable logistic regression of LOW on SMOKE.

```
.  logit low smoke

Iteration 0:   Log Likelihood =  -117.336
Iteration 1:   Log Likelihood = -114.9123
Iteration 2:   Log Likelihood = -114.9023

Logit Estimates                               Number of obs  =       189
                                              chi2(1)        =      4.87
                                              Prob > chi2    =    0.0274
Log Likelihood =  -114.9023                   Pseudo R2      =    0.0207


------------------------------------------------------------------------------
     low |     Coef.    Std. Err.       z      P>|z|     [95% Conf. Interval]
---------+--------------------------------------------------------------------
   smoke |   .7040592    .3196386     2.203    0.028     .0775791     1.330539
   _cons |  -1.087051    .2147299    -5.062    0.000    -1.507914    -.6661886
------------------------------------------------------------------------------
```

$$P_0 = \frac{e^{\beta_0}}{1+e^{\beta_0}} = \frac{e^{-1.087051}}{1+e^{-1.087051}} = .2522$$

π is the fraction of subjects in the study expected to have $x = 0$. The value of $\pi = 0.6085$ can be seen in the cross-classification shown above.

Values for $z_{1-\alpha}$ and $z_{1-\theta}$ are determined by the values of α and θ specified in the problem. In this problem, $\alpha = 0.05$ and $\theta = 0.20$ resulting in $z_{1-\alpha} = 1.645$ and $z_{1-\theta} = 0.842$.

We wish to calculate the sample size necessary to detect an odds ratio of 2.5, therefore the value of β_1^* is the ln(2.5).

These values can be substituted into Equation 8.44:

$$n = \left(1 + 2(0.25)\right) * \frac{\left(1.645\sqrt{\dfrac{1}{1-0.6085} + \dfrac{1}{0.6085}} + 0.842\sqrt{\dfrac{1}{1-0.6085} + \dfrac{1}{0.6085 * e^{\ln(2.5)}}}\right)^2}{(0.25)(\ln[2.5])^2}$$

When simplified the above equation indicates that a sample size of 171 is necessary to detect that the odds of a low birth weight baby among women who smoke during pregnancy is 2.5 times that of women who do not smoke, using a 5 percent type I error probability and 80 percent power.

12. *Consider the low birth weight study. What sample size would be needed in a new study to be able to detect that the odds of a low birthweight baby decrease at a rate of 10 percent per 10 pound increase in weight at the last menstrual period, using a 5 percent type I error probability and 80 percent power?*

Use Equation 8.45 to calculate the necessary sample size.

$$n = \left(1 + 2P_0\delta\right) * \frac{\left(z_{1-\alpha} + z_{1-\theta}e^{-0.25\beta_1^{*2}}\right)^2}{P_0\beta_1^{*2}}$$

where,

P_0 is the value of the logistic probability evaluated at the mean of the standardized covariates. First the covariate LWT must be standardized. This is done by subtracting the mean value of LWT from each individual's value of LWT and dividing the difference by the standard deviation of LWT. The necessary output is shown below.

```
. sum lwt

    Variable |     Obs      Mean   Std. Dev.      Min      Max
-------------+--------------------------------------------------
         lwt |     189   129.8148   30.57938       80      250
```

```
. gen lwtst=(lwt-129.8148)/30.57938
```

```
. logit low lwtst

Iteration 0:   Log Likelihood =  -117.336
Iteration 1:   Log Likelihood =-114.41626
Iteration 2:   Log Likelihood =-114.34546
Iteration 3:   Log Likelihood =-114.34533

Logit Estimates                                    Number of obs =      189
                                                   chi2 (1)      =     5.98
                                                   Prob > chi2   = 0.0145
Log Likelihood = -114.34533                        Pseudo R2     = 0.0255

---------------------------------------------------------------------------
    low |     Coef.   Std. Err.       z     P>|z|     [95% Conf. Interval]
--------+------------------------------------------------------------------
  lwtst |  -.4298929   .1886616    -2.279   0.023     -.7996628    -.060123
  _cons |   -.826656   .1625129    -5.087   0.000     -1.145175   -.5081366
---------------------------------------------------------------------------
```

P_0 can be obtained using Equation 8.47 and the results from the univariable logistic regression of LOW on LWTST.

$$P_0 = \frac{e^{\beta_0}}{1 + e^{\beta_0}} = \frac{e^{-0.826656}}{1 + e^{-0.826656}} = 0.3044$$

Values for $z_{1-\alpha}$ and $z_{1-\theta}$ are determined by the values of α and β specified in the problem. In this problem, $\alpha = 0.05$ and $\theta = 0.20$ resulting in $z_{1-\alpha} = 1.645$ and $z_{1-\theta} = 0.842$.

We wish to calculate the sample size necessary to detect an odds ratio of 0.9 for a 10 pound increase in weight at the last menstrual period. Since the variable LWT has been standardized, the sample size will reflect the odds for a one-unit change in the standard deviation of the variable. The standard deviation of LWT is roughly 30. The appropriate value for β_1^* can be obtained by calculating the value for the odds ratio corresponding to a 30 pound increase in weight at the last menstrual period. This can be done by taking the log of 0.9 and dividing by 10 to obtain "raw" coefficient (for a one-pound change in LWT) and multiplying by 30. When this value is exponentiated it shows that the odds of a low birthweight baby among women who are 30 pounds heavier are 0.729 times that of women whose weight is at the reference value.

$$\beta_1^* = \frac{\ln(0.9)}{10} * 30 = -0.3161$$

$$OR_{30 \text{ pound increase}} = e^{-0.3161} = 0.729$$

The value of δ can be calculated using Equation 8.46:

$$\delta = \frac{1 + \left(1 + \beta_1^{*2}\right)e^{1.25\beta_1^{*2}}}{1 + e^{-0.25\beta_1^{*2}}} = \frac{1 + \left(1 + [\ln(0.729)]^2 \, e^{1.25[\ln(0.729)]^2}\right)^2}{1 + e^{-0.25[\ln(0.729)]^2}} = 1.1336$$

These values can be substituted into Equation 8.45:

$$n = \left(1 + 2(0.3044)(1.1336)\right) * \frac{\left(1.645 + 0.842e^{-0.25[\ln(0.729)]^2}\right)^2}{(0.3044)(\ln[0.729])^2}$$

When simplified the above equation indicates that a sample size of 339 is necessary to detect that the odds of a low birth weight baby decrease at a rate of 10 percent per 10 pound increase in weight at the last menstrual period, using a 5 percent type I error probability and 80 percent power.

13. *Repeat Exercise 11 assuming that you plan to use a model that contains age, weight of the mother at the last menstrual period and race.*

    Use Equation 8.49 to calculate the necessary sample size.

$$n = \frac{\left(1 + 2P_0\right)}{1 - \rho^2} * \frac{\left(z_{1-\alpha}\sqrt{\dfrac{1}{1-\pi} + \dfrac{1}{\pi}} + z_{1-\theta}\sqrt{\dfrac{1}{1-\pi} + \dfrac{1}{\pi e^{\beta_1^*}}}\right)^2}{P_0 \beta_1^{*2}}$$

where,

    P_0 can be obtained using Equation 8.47 and the results from the multivariable logistic regression of LOW on SMOKE, AGEST, LWTST and RACE. AGEST and LWTST are the standardized values of AGE and LWT.

```
. logit low agest lwtst race_2 race_3 smoke

Iteration 0:   Log Likelihood =  -117.336
Iteration 1:   Log Likelihood =-107.59043
Iteration 2:   Log Likelihood =-107.29007
Iteration 3:   Log Likelihood =-107.28862
Iteration 4:   Log Likelihood =-107.28862

Logit Estimates                               Number of obs =      189
                                              chi2(5)       =    20.09
                                              Prob > chi2   =   0.0012
Log Likelihood = -107.28862                   Pseudo R2     =   0.0856


---------------------------------------------------------------------------
    low |     Coef.   Std. Err.       z      P>|z|     [95% Conf. Interval]
--------+------------------------------------------------------------------
  agest |  -.1191052   .1810584    -0.658    0.511    -.4739732    .2357629
  lwtst |  -.3830271   .1952749    -1.961    0.050    -.7657587   -.0002954
 race_2 |   1.231671   .5171518     2.382    0.017     .2180725    2.24527
 race_3 |   .9432627   .4162322     2.266    0.023     .1274626    1.759063
  smoke |   1.054439   .3799999     2.775    0.006     .3096526    1.799225
  _cons |  -1.815918   .3658892    -4.963    0.000    -2.533047   -1.098788
---------------------------------------------------------------------------
```

$$P_0 = \frac{e^{\beta_0}}{1 + e^{\beta_0}} = \frac{e^{-1.815918}}{1 + e^{-1.815918}} = 0.1399$$

π is the fraction of subjects in the study expected to have $x = 0$. The value of $\pi = 0.6085$ was calculated in problem 11.

Values for $z_{1-\alpha}$ and $z_{1-\theta}$ are determined by the values of α and β specified in the problem. In this problem, $\alpha = 0.05$ and $\beta = 0.20$ resulting in $z_{1-\alpha} = 1.645$ and $z_{1-\theta} = 0.842$.

We wish to calculate the sample size necessary to detect an odds ratio of 2.5, therefore the value of β_1^* is the $\ln(2.5)$.

The value of ρ^2 can be obtained by finding the squared correlation between the values of the dichotomous covariate (SMOKE) and the fitted values from a logistic regression of SMOKE on all other variables in the model. This can be obtained using Equation 5.6. The value of ρ^2 is shown below.

$$\rho^2 = 0.1444$$

These values can be substituted into Equation 8.49:

$$n = \frac{(1 + 2(0.1399))}{1 - 0.1444} * \frac{\left(1.645 \sqrt{\dfrac{1}{1 - 0.6085} + \dfrac{1}{0.6085}} + 0.842 \sqrt{\dfrac{1}{1 - 0.6085} + \dfrac{1}{0.6085 * e^{\ln(2.5)}}} \right)^2}{(0.1399)(\ln[2.5])^2}$$

When simplified the above equation indicates that a sample size of 304 is necessary to detect that the odds of a low birth weight baby among women who smoke during pregnancy is 2.5 times that of women who do not smoke, using a model that contains age, weight of the mother at the last menstrual period and race, using a 5 percent type I error probability and 80 percent power.

14. *Repeat Exercise 12 assuming that you plan to use a model that contains age, smoking status during pregnancy and race.*

Use Equation 8.48 to calculate the necessary sample size.

$$n = \frac{(1 + 2P_0\delta)}{(1 - \rho^2)} * \frac{\left(z_{1-\alpha} + z_{1-\theta} e^{-0.25\beta_1^{*2}} \right)^2}{P_0 \beta_1^{*2}}$$

where,

P_0 can be obtained using Equation 8.47 and the results from the multivariable logistic regression of LOW on SMOKE, AGEST, LWTST and RACE. AGEST and LWTST are the standardized values of AGE and LWT.

```
. logit low agest lwtst race_2 race_3 smoke

Iteration 0:   Log Likelihood =  -117.336
Iteration 1:   Log Likelihood =-107.59043
Iteration 2:   Log Likelihood =-107.29007
Iteration 3:   Log Likelihood =-107.28862
Iteration 4:   Log Likelihood =-107.28862

Logit Estimates                          Number of obs =      189
                                         chi2(5)       =    20.09
                                         Prob > chi2   = 0.0012
Log Likelihood = -107.28862              Pseudo R2     = 0.0856


---------------------------------------------------------------------
    low |     Coef.   Std. Err.      z     P>|z|    [95% Conf. Interval]
--------+------------------------------------------------------------
  agest | -.1191052  .1810584    -0.658   0.511   -.4739732   .2357629
  lwtst | -.3830271  .1952749    -1.961   0.050   -.7657587  -.0002954
 race_2 |  1.231671  .5171518     2.382   0.017    .2180725    2.24527
 race_3 |  .9432627  .4162322     2.266   0.023    .1274626   1.759063
  smoke |  1.054439  .3799999     2.775   0.006    .3096526   1.799225
  _cons | -1.815918  .3658892    -4.963   0.000   -2.533047  -1.098788
---------------------------------------------------------------------
```

$$P_0 = \frac{e^{\beta_0}}{1+e^{\beta_0}} = \frac{e^{-1.815918}}{1+e^{-1.815918}} = 0.1399$$

Values for $z_{1-\alpha}$ and $z_{1-\theta}$ are determined by the values of α and θ specified in the problem. In this problem, $\alpha = 0.05$ and $\theta = 0.20$ resulting in $z_{1-\alpha} = 1.645$ and $z_{1-\theta} = 0.842$.

We wish to calculate the sample size necessary to detect an odds ratio of 0.9 for a 10 pound increase in weight at the last menstrual period. Since the variable LWT has been standardized, the sample size will reflect the odds for a one-unit change in the standard deviation of the variable. The standard deviation of LWT is roughly 30. The appropriate value for β_1^* can be obtained by calculating the value for the odds ratio corresponding to a 30 pound increase in weight at the last menstrual period. This can be done by taking the log of 0.9 and dividing by 10 to obtain "raw" coefficient (for a one-pound change in LWT) and multiplying by 30. When this value is exponentiated it shows that the odds of a low birthweight baby among women who are 30 pounds heavier are 0.729 times that of women whose weight is at the reference value.

$$\beta_1^* = \frac{\ln(0.9)}{10} * 30 = -0.3161$$

$$OR_{30\ pound\ increase} = e^{-0.3161} = 0.729$$

The value of δ can be calculated using Equation 8.46:

$$\delta = \frac{1 + \left(1 + \beta_1^{*2}\right)e^{1.25\beta_1^{*2}}}{1 + e^{-0.25\beta_1^{*2}}} = \frac{1 + \left(1 + \left[\ln(0.729)\right]^2 e^{1.25\left[\ln(0.729)\right]^2}\right)^2}{1 + e^{-0.25\left[\ln(0.729)\right]^2}} = 1.1336$$

The value of ρ^2 can be obtained by using a multiple linear regression package with LWTST as the dependent variable and the remaining variables as covariates. The value of ρ^2 shown below is 0.1238.

```
. fit lwtst agest race_2 race_3 smoke

  Source |       SS         df        MS                Number of obs =      189
---------+------------------------------               F(  4,   184) =     6.50
   Model |  23.2829199       4    5.82072998            Prob > F      =   0.0001
Residual |  164.717085     184     .895201549           R-squared     =   0.1238
---------+------------------------------               Adj R-squared =   0.1048
   Total |  188.000005     188    1.00000003            Root MSE      =   .94615


---------------------------------------------------------------------------------
   lwtst |     Coef.    Std. Err.       t      P>|t|      [95% Conf. Interval]
---------+-----------------------------------------------------------------------
   agest |   .1768391    .0709888     2.491    0.014      .0367824     .3168958
  race_2 |   .5431263    .2141709     2.536    0.012      .1205799     .9656726
  race_3 |  -.4023499    .1631948    -2.465    0.015     -.7243236    -.0803762
   smoke |  -.1992618    .1508143    -1.321    0.188     -.4968094     .0982858
   _cons |   .1459347    .1282604     1.138    0.257     -.1071155     .3989848
---------------------------------------------------------------------------------
```

These values can be substituted into Equation 8.48:

$$n = \frac{\left(1 + 2(0.1399)(1.1336)\right)}{1 - 0.1238} * \frac{\left(1.645 + 0.842e^{-0.25\left[\ln(0.729)\right]^2}\right)^2}{(0.1399)\left(\ln[0.729]\right)^2}$$

When simplified the above equation indicates that a sample size of 655 is necessary to detect that the odds of a low birth weight baby decrease at a rate of 10 percent per 10 pound increase in weight at the last menstrual period, using a model containing age, smoking status during pregnancy and race, using a 5 percent type I error probability and 80 percent power.

15. *If the sample size obtained in Exercisess 13 or 14 is larger than the original study size of 189, use the suggested method for obtaining a larger study to explore the effect the larger study size has on estimated coefficients, standard errors and confidence intervals.*

In Exercise 13, the sample size calculations suggest a sample size of 304, which is larger than the original study size of 189. One way to assess the precision obtained from modeling with a sample of 304 subjects is to construct a pseudo-study by generated a weight for each of the 189 original subjects. The weight for each subject will be the ideal sample size (304) divided by the original sample size (189). This corresponds to a weight of roughly 1.608 for each subject. A weighted logistic regression can be performed and the estimated coefficients, standard errors and confidence intervals can be compared with the results from the unweighted regression.

```
. gen wt1=304/189

. xi:logit low smoke age lwt i.race [iweight=wt1]
i.race            _Irace_1-3        (naturally coded; _Irace_1 omitted)

likelihood = -172.57005

Logit estimates                          Number of obs   =        304
                                         LR chi2(5)      =      32.32
                                         Prob > chi2     =     0.0000
Log likelihood = -172.57005              Pseudo R2       =     0.0856

------------------------------------------------------------------------
     low |     Coef.   Std. Err.      z    P>|z|    [95% Conf. Interval]
---------+--------------------------------------------------------------
   smoke |  1.054439   .2996247    3.52   0.000    .4671851    1.641692
     age | -.0224783    .026943   -0.83   0.404   -.0752855     .030329
     lwt | -.0125257   .0050351   -2.49   0.013   -.0223944    -.002657
 _Irace_2 |  1.231671   .4077671    3.02   0.003    .4324626     2.03088
 _Irace_3 |  .9432627   .3281933    2.87   0.004    .3000156    1.58651
    _cons |  .3324516    .873385    0.38   0.703   -1.379351    2.044255
------------------------------------------------------------------------
--
```

```
. logit low smoke age lwt race_2 race_3

Logit Estimates                          Number of obs =        189
                                         chi2(5)       =      20.09
                                         Prob > chi2   =     0.0012
Log Likelihood = -107.28862              Pseudo R2     =     0.0856

------------------------------------------------------------------------
     low |     Coef.   Std. Err.      z    P>|z|    [95% Conf. Interval]
---------+--------------------------------------------------------------
   smoke |  1.054439   .3799999   2.775   0.006    .3096526    1.799225
     age | -.0224783   .0341705  -0.658   0.511   -.0894512    .0444947
     lwt | -.0125257   .0063858  -1.961   0.050   -.0250417   -9.66e-06
  race_2 |  1.231671   .5171518   2.382   0.017    .2180725     2.24527
  race_3 |  .9432627   .4162322   2.266   0.023    .1274626    1.759063
    _cons |  .3324516   1.107673   0.300   0.764   -1.838548    2.503451
------------------------------------------------------------------------
```

Results from the pseudo-study show that the coefficients in the model do not change. Estimated standard errors are smaller in the weighted analysis, by a factor of $\left(1/\sqrt{wt1}\right)$ reflecting the larger sample size. For example, in the unweighted analysis smoke has $\hat{SE} = 0.379999$ and in the weighted analysis $\hat{SE} = 0.22996347 = 0.379999 \times \left(1//\sqrt{1.608}\right)$. We note that as a result the 95% confidence intervals for the estimated coefficients are narrower.

The same approach can be used for the sample size suggested in problem 14. In this problem, the sample size calculations suggest a sample size of 655, which is larger than the original study size of 189. One way to assess the precision obtained from modeling with a sample of 655 subjects is to construct a pseudo-study by generated a weight for each of the 189 original subjects. The weight for each subject will be the ideal sample size (655) divided by the original sample size (189). This corresponds to a weight of roughly 3.47 for each subject. A weighted logistic regression can be performed and the estimated coefficients, standard errors and confidence intervals can be compared with the results from the unweighted regression.

```
. gen wt2=655/189

xi: logit low lwt smoke age i.race [iweight=wt2]
i.race              _Irace_1-3        (naturally coded; _Irace_1 omitted)


Logit estimates                              Number of obs   =        655
                                             LR chi2(5)      =      69.64
                                             Prob > chi2     =     0.0000
Log likelihood = -371.82033                  Pseudo R2       =     0.0856


--------------------------------------------------------------------------
        low |     Coef.   Std. Err.       z     P>|z|    [95% Conf. Interval]
------------+-------------------------------------------------------------
        lwt | -.0125257   .0034303    -3.65    0.000   -.0192489   -.0058025
      smoke |  1.054439   .2041238     5.17    0.000    .6543633    1.454514
        age | -.0224783   .0183553    -1.22    0.221    -.058454    .0134975
    _Irace_2 |  1.231671   .2777975     4.43    0.000    .6871983    1.776144
    _Irace_3 |  .9432627   .2235867     4.22    0.000    .5050409    1.381484
       _cons |  .3324516   .5950067     0.56    0.576   -.8337401    1.498643
--------------------------------------------------------------------------
```

```
. logit low   lwt smoke age race_2 race_3


Logit Estimates                              Number of obs =        189
                                             chi2(5)         =      20.09
                                             Prob > chi2     =     0.0012
Log Likelihood = -107.28862                  Pseudo R2       =     0.0856


--------------------------------------------------------------------------
        low |     Coef.   Std. Err.       z     P>|z|    [95% Conf. Interval]
----------+---------------------------------------------------------------
        lwt | -.0125257   .0063858   -1.961    0.050   -.0250417   -9.66e-06
      smoke |  1.054439   .3799999    2.775    0.006    .3096526    1.799225
        age | -.0224783   .0341705   -0.658    0.511   -.0894512    .0444947
     race_2 |  1.231671   .5171518    2.382    0.017    .2180725     2.24527
     race_3 |  .9432627   .4162322    2.266    0.023    .1274626    1.759063
      _cons |  .3324516   1.107673    0.300    0.764   -1.838548    2.503451
--------------------------------------------------------------------------
```

As before, results from the pseudo-study show that the coefficients in the model do not change. Estimated standard errors are smaller in the weighted analysis by a factor of $\left(1/\sqrt{wt2}\right)$ reflecting the larger sample size. For example, in the unweighted analysis smoke has $\hat{SE} = 0.379999$ and in the weighted analysis $\hat{SE} = 0.2041238 = 0.379999 \times \left(1/\sqrt{3.466}\right)$. We note that as a result the 95% confidence intervals for the estimated coefficients are narrower.